T0313368

Advanced VLSI Technology
Technical Questions and Solutions

RIVER PUBLISHERS SERIES IN CIRCUITS AND SYSTEMS

Series Editors:

MASSIMO ALIOTO
National University of Singapore
Singapore

KOFI MAKINWA
Delft University of Technology
The Netherlands

DENNIS SYLVESTER
University of Michigan
USA

Indexing: All books published in this series are submitted to the Web of Science Book Citation Index (BkCI), to SCOPUS, to CrossRef and to Google Scholar for evaluation and indexing.

The "River Publishers Series in Circuits and Systems" is a series of comprehensive academic and professional books which focus on theory and applications of Circuit and Systems. This includes analog and digital integrated circuits, memory technologies, system-on-chip and processor design. The series also includes books on electronic design automation and design methodology, as well as computer aided design tools.

Books published in the series include research monographs, edited volumes, handbooks and textbooks. The books provide professionals, researchers, educators, and advanced students in the field with an invaluable insight into the latest research and developments.

Topics covered in the series include, but are by no means restricted to the following:

- Analog Integrated Circuits
- Digital Integrated Circuits
- Data Converters
- Processor Architecures
- System-on-Chip
- Memory Design
- Electronic Design Automation

For a list of other books in this series, visit www.riverpublishers.com

Advanced VLSI Technology
Technical Questions and Solutions

Cherry Bhargava

Lovely Professional University
India

Gaurav Mani Khanal

University of Rome Tor Vergata
Italy

River Publishers

Routledge
Taylor & Francis Group

LONDON AND NEW YORK

Published 2020 by River Publishers
River Publishers
Alsbjergvej 10, 9260 Gistrup, Denmark
www.riverpublishers.com

Distributed exclusively by Routledge
4 Park Square, Milton Park, Abingdon, Oxon OX14 4RN
605 Third Avenue, New York, NY 10017, USA

Advanced VLSI Technology Technical Questions and Solutions / by Cherry Bhargava, Gaurav Mani Khanal.

Routledge is an imprint of the Taylor & Francis Group, an informa business

ISBN 978-87-7022-174-0 (print)

While every effort is made to provide dependable information, the publisher, authors, and editors cannot be held responsible for any errors or omissions.

This Book is dedicated to

My Parents, Husband
and
Loving Daughters Mishty & Mauli

— Cherry Bhargava

My Parents, Wife
and Daughter Gaurisha

— Gaurav Mani Khanal

Contents

Preface

As technology advances, the need for low cost and reliable systems is enhancing exponentially. The rapid development in semiconductor technology is leading to low cost tiny and fast systems that have the capability of a supercomputer. The design of such small and efficient systems is now witnessed in every field like a toy to the satellite. In digital design, you can create fast and powerful circuits in smaller and smaller devices such as ASIC-Application specific integrated circuits (ASICs) and Systems-on-a-chip (SoCs) which are highly complex mixed-signal circuits (Digital and Analog on the same chip) using various EDA (Electronic Design Automation) tools.

The VLSI technology is shining brighter than the sun and demand of VLSI engineers is always in high demand. VLSI is a broad spectrum of technologies and there are several subcategories of jobs that companies hire for which can be broadly in three categories such as front end design (RTL design, microarchitecture, Functional Verification, Synthesis, etc), back end design (Floor-planning, Placement and Routing, Timing and Clock Tree synthesis, etc) as well as in Silicon Validation and Testing (hardware and software framework and test generation for silicon testing in lab). The purpose of writing this book is to demystify the hardware design for software engineers as well as to elucidate the VLSI design and related technologies for students as well as researchers. This book refreshes the basic concepts for designing and developing an integrated circuit using advanced tools and techniques.

Organization of the Book

This book is divided into 4 units, which cover all the subdomains of advanced VLSI design. After summarising the latest and advanced concepts of related technology, the technical questions related to that particular technology is given with solutions.

Unit 1: Focuses on the validation of static timing performance of a design by introducing various timing paths and violations. The concept of clocking and metastability is discussed in brief along with interview questions.

Unit 2: Gives a glimpse of various layout and design rules for CMOS based technology. The rules of the stick diagram are described in short along with probable interview questions.

Unit 3: Explains how to place and route the various components on a single chip, which will optimize the design parameters.

Unit 4: Deals with testing of VLSI chip using various test patterns and techniques. The stuck-at faults are described along with the concept of controllability.

Appendix A: List of Digital circuit IC numbers.

Appendix B: List of Keywords, System Tasks & Compiler Directives Used in Verilog HDL.

Acknowledgement

At this movement of our substantial enhancement, before we get into the thick of the things, we would like to add a few heartfelt words for the people who gave their unending support with their unfair humor and warm wishes. First and foremost, praises and thanks to the God, the Almighty, for his showers of blessings throughout, to complete this book successfully.

We want to acknowledge our students who provided us with the impetus to write a more suitable text. Our supporting family members deserve great acknowledgement in true sense who have always been a force to keep me riveted to our dedication towards the present book.

Besides, we thank all our friends, well-wishers, respondents and academicians who helped throughout my journey from inception to completion.

List of Figures

List of Tables

List of Abbreviations

AC	Alternating Current
ASIC	Application Specific Integrated Circuit
ATE	Automated Test Equipment
ATPG	Automatic Test Pattern Generation
BiCMOS	Bipolar Complementary Metal-Oxide Semiconductor logic
BIST	Built-In Self Test
BJT	Bipolar Junction Transistor
CAD	Computer-Aided Design
CCS	Composite Current Source
CDMA	Code Division Multiple Access
CMOS	Complementary Metal Oxide Semiconductor logic
CMP	Chemical Mechanical Polishing
CMRR	Common Mode Rejection Ratio
CNTFET	Carbon NanoTube Field-Effect Transistors
CRO	Cathode Ray Oscilloscope
CTO	Clock Tree Optimization
CTS	Clock Tree Synthesis
CUT	Circuit Under Test
DC	Direct Current
DFM	Defects Per Million
DFT	Design For Test
DRC	Design Rule Check
DRC	Design Rule Check
DSO	Digital Storage Oscilloscope
DUT	Device Under Test
EDA	Electronic Design Automation
EM	ElectroMigration
ERC	Electric Rule Check
FET	Field-Effect Transistors
FIFO	First In First Out
FinFET	Fin Field Effect Transistor

FIT	Failure In Time
FPGA	Field Programmable Gate Array
HDL	Hardware Description Language
HFNS	High Fanout Net Synthesis
IC	Integrated Circuit
ICG	Integrated Clock Gating
IGBT	Insulated Gate Bipolar Transistor
LDR	Light Dependent Resistor
LEC	Logical Equivalence Check
LED	Light Emitting Diode
LVDT	Linear Variable Differential Transformer
LVS	Layout versus Schematic
LVT	Low Threshold Voltage
MIGFET	Multiple-Independent-Gate Field-Effect Transistor
MMMC	Multi-Mode Multi-Corner
MTBF	Mean Time Between Failure
MTTF	Mean Time To Failure
MTTR	Mean Time To Repair
MuGFET	Multiple-Gate Field-Effect Transistor
NDM	Nonlinear Delay Model
NDR	Non-Default Rules
NMOS	N-type Metal Oxide Semiconductor logic
OCV	On-Chip Variation
P&R	Placement & Routing
PDA	Physical Design Automation
PLL	Phase-Locked Loops
PMOS	P-type Metal Oxide Semiconductor logic
PV	PhotoVoltaic
PVT	Process Voltage Temperature
QCA	Quantum-Dot Cellular Automata
QL	Quality Level
RTL	Register Transfer Level
SET	Single Electron Transistor
SOA	Safe Operating Area
SoC	System On Chip
SOI	Silicon On Insulator
STA	Static Timing Analysis
SVT	Standard Threshold Voltage
TDMA	Time Division Multiple Access

TIE	Time Interval Error
VHDL	Very High-Speed Integrated Circuit Hardware Description Language
VLSI	Very Large Scale Integration
WLM	Wire Load Model

1

Static Timings Analysis

Static timing analysis (STA) is a simulation method of computing the expected timing of a digital circuit without requiring simulation of the full circuit.

High-performance integrated circuits have traditionally been characterized by the clock frequency at which they operate. Measuring the ability of a circuit to operate at the specified speed requires an ability to measure, during the design process, its delay at numerous steps. Static timing analysis plays a vital role in facilitating the fast and reasonably accurate measurement of circuit timing.

1.1 Timing Components

Timing components are one of the most ubiquitous components in electronics. They are needed in nearly every complex design and all our electronics would not work without them. The basic clock/timing components are:

1.1.1 Clock Signal

In electronics and especially synchronous digital circuits, a clock signal is a particular type of signal that oscillates between a high and a low state and is used as a metronome to coordinate actions of digital circuits. A clock signal is produced by a clock generator.

There are different ways a clock signal can to be produced, but they all start with the crystal resonator. A crystal resonator is commonly referred to as a crystal. To operate, crystals are combined with an amplifier circuit to apply voltage to an electrode near or on the crystal.

A clock signal is a signal which is used to trigger sequential devices.

1.1.2 Quartz Crystal

The quartz crystal is a tiny slit of quartz with each of the two surfaces metalized and attached with an electrical connection. The physical size and shape of the quartz crystal must be precisely cut because this determines the frequency of oscillations produced from the crystal.

Once the crystal is cut and shaped, it cannot be used at any other frequency. Quartz crystals are more commonly used since the frequency generated from quartz crystals is more resistant to changes in temperature. If an internal RC resonator was used instead, changes in temperature would affect the behavior of the oscillator, leading to changes in the output frequency.

1.1.3 Crystal Oscillator

A crystal oscillator is an electronic oscillator circuit that uses the mechanical resonance of a vibrating crystal of piezoelectric material to create an electrical signal with a precise frequency.

Crystals have a sinusoidal output and are typically used if the target IC has an integrated oscillator and on-chip phase-locked loops (PLLs) for internal timing. When crystal and oscillation circuits are combined in the same package, it is commonly referred to as a crystal oscillator. This quartz piezo-electric oscillator outputs a usable oscillating signal, most commonly a square wave with a 50% duty cycle. Usually, this clock signal is fixed at a constant frequency and synchronization may become active at either the rising or falling edge of each clock cycle.

1.1.4 Clock Generator

A clock generator is an electronic oscillator (circuit) that produces a clock signal for use in synchronizing a circuit's operation. A clock generator combines an oscillator with one or more PLLs, output dividers, and output buffers. Clock generators and clock buffers are useful when several frequencies are required and the target ICs are all on the same board or in the same FPGA.

In some applications, FPGA/ASICs have multiple time domains for the data path, control plane, and memory controller interface, and as a result, require multiple unique reference frequencies. In most cases, the oscillator is external to the clock generator, although it is becoming more common that oscillators are combined into the same package as the clock generator to consolidate the bill of material cost and complexity, alongside other advantages.

There are many different types of clock generators and each is optimized for different performance and cost targets depending on the application.

1.1.5 Clock Rate

The clock rate typically refers to the frequency at which the clock generator of a processor can generate pulses, which are used to synchronize the operations of its components and is used as an indicator of the processor's speed. It is measured in clock cycles per second or its equivalent, the SI unit hertz (Hz).

1.1.6 Clock Multiplier

Many modern microcomputers use a "clock multiplier" which multiplies a lower frequency external clock to the appropriate clock rate of the microprocessor. This allows the CPU to operate at a much higher frequency than the rest of the computer, which affords performance gains in situations where the CPU does not need to wait on an external factor (like memory or input/output).

1.1.7 Clock Tree

The clock distribution network (or clock tree, when this network forms a tree) distributes the clock signal(s) from a common point to all the elements that need it.

The distribution of clock in design is known as a clock tree. The clock signal should be reached to every element at the almost same time.

1.1.8 Clock Phase

Most integrated circuits (ICs) of sufficient complexity use a clock signal to synchronize different parts of the circuit, cycling at a rate slower than the worst-case internal propagation delays. In some cases, more than one clock cycle is required to perform a predictable action. As ICs become more complex, the problem of supplying accurate and synchronized clocks to all the circuits becomes increasingly difficult. The preeminent example of such complex chips is the microprocessor, the central component of modern computers, which relies on a clock from a crystal oscillator.

- Single-phase clock

All clock signals are (effectively) transmitted on 1 wire.

• Two-phase clock

In synchronous circuits, a "two-phase clock" refers to clock signals distributed on 2 wires, each with non-overlapping pulses. Traditionally one wire is called "phase 1" or "$\varphi1$", the other wire carries the "phase 2" or "$\varphi2$" signal.

• Four-phase clock

Some early integrated circuits use four-phase logic, requiring a four-phase clock input consisting of four separate, non-overlapping clock signals.

1.1.9 Clock Gating

Clock gating is a popular technique used in many synchronous circuits for reducing dynamic power dissipation. Clock gating saves power by adding more logic to a circuit to prune the clock tree. Pruning the clock disables portions of the circuitry so that the flip-flops in them do not have to switch states. Switching states consume power. When not being switched, the switching power consumption goes to zero, and only leakage currents are incurred.

Clock gating logic can be added into a design in a variety of ways:

- Coded into the register transfer level (RTL) code as enable conditions that can be automatically translated into clock gating logic by synthesis tools (fine grain clock gating).
- Inserted into the design manually by the RTL designers (typically as module-level clock gating) by instantiating library-specific integrated clock gating (ICG) cells to gate the clocks of specific modules or registers.
- Semi-automatically inserted into the RTL by automated clock gating tools. These tools either insert ICG cells into the RTL or add enable conditions into the RTL code. These typically also offer sequential clock gating optimizations.

1.1.10 Clock Jitter

Clock jitter is a characteristic of the clock source and the clock signal environment. It can be defined as a "deviation of a clock edge from its ideal location." Clock jitter is typically caused by clock generator circuitry, noise, power supply variations, interference from nearby circuitry, etc.

Clock skew is two different flip flops receive the clock signal at slightly different times due to difference in clock net length but clock jitter is on the same flip flop but the position of clock edge moves edge to edge due to some noise in the oscillator. So, clock skew is the difference in arrival time of clock signals at different pins.

1.1.11 Clock Latency

Clock latency is the total delay that a clock signal takes to reach a sink or a destination pin, which typically is the clock pin of the flip-flops or the latches, from a clock source.

1.2 Crosstalk

Crosstalk is a phenomenon by which logic transmitted in a VLSI circuit or a net/wire creates an undesired effect on the neighboring circuit or nets/wires due to capacitive coupling.

1.2.1 Crosstalk Noise Due To Coupling Capacitance

Figure 1.1 depicts the crosstalk noise which leads to logic failures, here V is the victim and A is an aggressor. The disturbance at 'A' can potentially cause a disturbance at 'V', because of the mutual coupling capacitance, and if the disturbance at 'V' crosses noise threshold of the receiving gate 'R', then it may change the logic at the output of 'R' i.e., output of 'R', which is supposed to be at logic '1', might switch to logic '0', as it senses a logic '1' at its input, due to the noise-induced on its input by the disturbance at 'A'.

1.2.2 Coupling Capacitance

In deep sub-micron technology (i.e. <130 nm) and below, the lateral capacitance between nets/wires on silicon becomes much more dominant than the interlayer capacitance. Hence, there is a capacitive coupling between the nets, that can lead to logic failures and degradation of timing in VLSI circuits. The Figure 1.2 represents Synchronous circuits data transmission with delay.

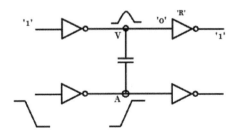

Figure 1.1   Crosstalk noise.

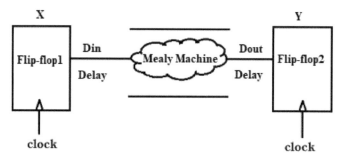

Figure 1.2   Synchronous circuits data transmission with delay.

1.3 Static Timing Analysis

Static Timing Analysis (STA) is one of the techniques to verify the design in terms of timing. This kind of analysis does not depend on any data or logic inputs, applied at the input pins. The input to an STA tool is the routed netlist, clock definitions (or clock frequency) and external environment definitions. The STA will validate whether the design could operate at the rated clock frequency, without any timing violations. Some of the basic timing violations are setup violation and hold violation.

Consider a flip-flop 'X' which generates data 'Din and it arrives as inputs to Mealy Machine after some delay q'(current state). Mealy Machine generates an output 'Dout', at q (next state). The receiver 'Y' receives 'Dout' after some delay. The same clock is used to control all transfer of data between flip-flops. For correct operation, this change in 'Dout' should be captured at the next clock edge by 'Y'.

Assume, the identical clock goes to both and there is some delay between Sender and Receiver. This delay will not be fixed as the effective load capacitance seen by each gate in the design is different. Other factors that affect the delay of a gate such as input transition, threshold voltage, drive strength, etc.

1.4 Unateness and Its Types

A function is said to be unate if the rise transition on the positive unate input variable causes the output to rise or no change and vice versa.

There are three types of Unate/Timing arc:

(a) Negative Unate

It means cell output logic is an inverted version of input logic. e.g. In inverter having input A and output Y, Y is -ve unate with respect to A.

(b) Positive Unate

It means cell output logic is as same as that of input.

(c) Non-Unate

If the clock sense is ambiguous

A clock signal is **positive unate** if a rising edge at the clock source can only cause a rising edge at the register clock pin, and a falling edge at the clock source can only cause a falling edge at the register clock pin.

A clock signal is **negative unate** if a rising edge at the clock source can only cause a falling edge at the register clock pin, and a falling edge at the clock source can only cause a rising edge at the register clock pin. In other words, the clock signal is inverted.

A clock signal is **not unate** if the clock sense is ambiguous as a result of non-unate timing arcs in the clock path. For example, a clock that passes through an XOR gate is not unate because there are nonunate arcs in the gate. The clocked sense could be either positive or negative, depending on the state of the other input to the XOR gate.

The positive and negative unateness are constraints defined in the library file and are defined for output pin to input pin.

Technical Questions with Solutions

- **Ques: 1. What do you mean by STA?**

Solution:

STA refers to Static Timing Analysis. It is a technique that is used to verify the design of a circuit in terms of timing. It validates whether the design could operate at the rated clock frequency. STA further checks all possible ways of timing violation.

- **Ques: 2. Why timing analysis is an important factor?**

Solution:

1. Timing analysis is used to select appropriate components, as few components are slow. The performance of the circuit degrades because the slow component introduces wait for the state. The fast component is costly. So, timing analysis selects the appropriate component as per the specific application.

2. Timing analysis verifies whether the circuit is properly designed and work with reliable output for all combinations of input.

• **Ques: 3. How many types of timing analysis are done in VLSI?**

Solution:

Timing analysis is of two types:

1. STA
2. DTA

STA: Static Timing Analysis: checks static delay requirement of circuit. It does not require any input or output variables.

DTA: Dynamic Timing Analysis: The function of DTA is to verify the design functionality with the help of input and output variables.

• **Ques: 4. What are the important features of STA?**

Solution:

1. No need for input-output variables.
2. Simple to use STA tools.
3. The input to STA is a library, netlist, constraints, and parasitic (R & C), all are commonly available.
4. STA uses device models based on lookup tables or constant I/V models. It uses the Elmore wire delay model.
5. STA performs worst-case analysis to check delay requirements of the circuit. It performs timing analysis on all possible paths i.e. it includes potential false paths also.
6. It is efficient for only a fully synchronous design.

So, the conclusion is:

• STA breaks the design into different timing paths.
• Calculate the signal propagation delay along each path.
• Check violation of timing constraints inside design and me/O interface.

• **Ques: 5. During timing analysis, what are the ideal characteristics of a clock?**

Solution:

1. The clock should be free of glitches.
2. The period of the clock should be properly defined and proper phase relationships should be established between two different clocks of interest.
3. The clock must meet pulse width requirements.

4. The Jitter parameter needs to be taken care of when the clock speeds increase. For example, PLL should have maximum jitter.
5. When data is transferred from one clock edge to another, the worst-case duty cycle should be used.

- **Ques: 6. What are the major functions of STA?**

Solution:

STA checks the following parameters:

1. Setup time
2. Hold time
3. Reset removal and reset recovery time
4. Clock gating
5. Min/max fan-out range
6. Maximum capacitance range
7. Clock pulse width requirements

- **Ques: 7. Which input files are required to run STA?**

Solution:

1. Gate level netlist
2. Parasitic files
3. Constraints
4. General setup scripts.

- **Ques: 8. When Static Timing Analysis is done?**

Solution:

STA can be done after synthesis. It should be done once before layout and 2–3 times after layout. The sign-off can be done after routing.

- **Ques: 9. How STA is different from circuit simulation?**

Solution:

1. As STA does not handle input-output variables so, it is faster than circuit simulation.
2. It provides deep insight by worst-case timing analysis of all possible logic conditions, whereas the circuit simulation verifies a particular set of input-output variables.

- **Ques: 10. How STA is performed on the circuit?**

Solution:

1. The circuit design is further bifurcated into a possible set of timing paths.

2. The signal propagation delay is calculated for all the paths.
3. STA tool analyses the timing constraint of all possible paths and compare with the ideal timing constraints and check whether there is any timing violation in the circuit.
4. It checks timing violations inside the design as well as at the input-output interface.

Design flow:
Simulation→Synthesis→STA→Layout→Sign-off

- **Ques: 11. For timing analysis, what are the various paths that the designer consider?**

Solution:

There are various types of paths, that are to be considered by designer:

1. Data path
2. Asynchronous path
3. Clock path
4. Clock gating path
5. The worst and best path
6. Capture and launch path
7. Critical path

- **Ques: 12. What do you mean by timing path? What are the start and endpoints?**

Solution:

For static timing analysis, various timing path and path delay is analyzed. The gate delays and net delays are used to calculate path delays. In the timing path, the data is launched (start point) and pass-through using combinational components and as soon as it meets with any sequential component (endpoint), it stops. If at both endpoints, there are sequential elements that are triggered by an asynchronous circuit i.e. using two different clocks, then for setup and hold time analysis, the LCM of both clock periods is considered. The launch and captured edge can be explored using LCM of the clock pulse.

- **Ques: 13. In the synchronous circuit, what is the first stage of timing delay?**

Solution:

In synchronous circuits, the timing path starts at the clock pin of Flip-flop A. The delay introduced from the clock edge to data output is known as the first stage of delay. The data goes through a series of combinational elements and interconnect wires. Each stage has a timing delay. When data reaches to another Flip-flop B, timing path stops. The clock divergence point is generated because the same clock is used to generate data through Flip-flop A and sample data through Flip-flop B. The Stages of timing delay is represented in Figure 1.3.

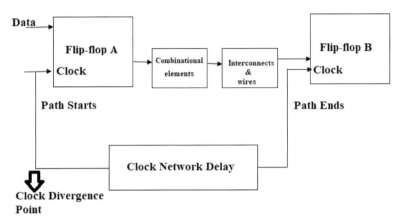

Figure 1.3 Stages of timing delay.

- **Ques: 14. What are the various timing paths that a designer to go through?**

Solution:

Following timing paths are majorly considered:

1. Clock pin of one register to D-pin of another register.
2. Input to D-pin of register.
3. D-pin of the register to output.
4. Input to output through combinational elements.
5. Input to the macro input pin, macro input to the macro output pin, macro output to the primary output pin.

- **Ques: 15. What do you mean by launch edge and capture edge?**

Solution:

In synchronous design, generation of data, certain computations, transfer of data, all are done within one clock cycle. At the rising or falling edge of the clock, the memory elements i.e. flip-flop A transfers the data from the input pin to the output pin. This active edge of the clock at which data is launched at the output of flip-flop A is also known as the launch edge.

The data needs to meet certain timing requirements before it reaches flip-flop B. At the next active clock edge, the data and computational results at the input pin of flip-flop B are captured and the data is transferred to the output pin of flip-flop B. This is known as capture edge.

- **Ques: 16. What do you mean by setup time and hold time?**

Solution:

The data needs to be settled before the capture edge of the clock activates. If the data does not settle before the capture edge, the flip-flop will enter into the metastability state. The time taken by input data to be stable before the capture edge of the clock is known as setup time.

When the capture edge of the clock is deactivated, the time for how long the data remains stable is known as hold time of flip-flop.

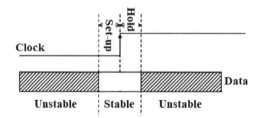

Figure 1.4   Setup and hold time.

This Figure 1.4 shows the setup time and hold time.

- **Ques: 17. Which factors decide setup time and hold time?**

Solution:

The set-up time and hold time are calculated by the input data slope, clock slope, and output load.

- **Ques: 18. What do you mean by setup time and hold time violation?**

Solution:

At the active edge of the clock, when the data is launched and transverse through flip-flop A and reaches output pin of flip-flop A with some delay. The data should be stable before the capture edge. But sometimes delay makes the circuit unstable and flip-flop enters into metastability and does not satisfy the set-up timing requirements. A similar condition is withheld time, thereafter the assertion of clock capture edge, the data becomes unstable, which violates the hold time requirement of flip-flop/sequential element. The hold time violations are functional failures.

• **Ques: 19. What are the main reasons for setup or hold time violations?**

Solution:

1. High clock slope
2. Very fast transition from the output of flip-flop A to the input of flip-flop B.
3. Sharp clock skew rate due to which second clock edge delays by a first clock edge. There is no synchronization in the alignment of two clock edges.
4. Capacitance coupling
5. Design issues

• **Ques: 20. What do you mean by critical path, false path, and multi-cycle path?**

Solution:

The static timing analysis tool is the exhaustive analysis tool that explores and analyses all the timing paths, even if it does not happen.

In timing analysis, the critical path is considered that timing-sensitive functional path which introduces the longest delay in the design. The timing path from the clock to the output of the first flip-flop may have some delay. Assuming both flip-flops are having the same clock if the delay (Clk-output of flip-flop A) is less than the clock period, it is known as timing requirement meets otherwise the timing requirement violates. The path with the highest delay is known as the critical path.

When no data is transferred from start to endpoint, this path is known as a false path. This is a functionally incorrect path. This path is intentionally inserted by the designer to develop a relation between asynchronous circuits. For example, in design, two D flip-flops are not enabled at the same time.

When the generation of data, transfer data, and computation of data takes place in more than one clock cycle, i.e. the data takes more than one cycle to travel from the start point to endpoint, is known as a multi cycling path.

• **Ques: 21. What is the worst path and best path?**

Solution:

In between the start point and the endpoint, there are many types of 'path'. The path which has the minimum delay is known as an early path, best path or minimum path i.e. through this path, the data takes minimum time to reach the endpoint. The path which is having the largest delay is known as the worst path, late path or maximum path, i.e. using this path, the data takes maximum time to reach the endpoint.

- **Ques: 22. Out of setup time violation and hold time violation, which is more dangerous to the design specifications and working mode?**

Solution:

The setup time violation is frequency-dependent. It can be removed or reduced by changing the frequency of the clock. Whereas the hold time violation is the functional failure of design. It is frequency independent. It cannot be repaired by slowing down the frequency of the clock, as it introduces data race. So, with a change in the clock period, the design with setup time violation can be used but the hold time violation persists in the design.

- **Ques: 23. What do you mean by the term "time borrowing"?**

Solution:

Time Borrowing is the concept of borrowing the time from the next clock cycle. It occurs in the case of the latch. It is also known as cycle stealing. It reduces the data time to arrive at the next clock cycle or in another case, it permits the design to use slack from the previous clock cycle.

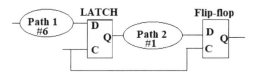

Figure 1.5   Time borrowing.

This example will clear the concept of time borrowing from the next cycle and slack from the previous clock cycle, as shown in Figure 1.5.

Assume that there is communication between two flip-flops, where the clock period is 5 ns. Path 1 has a timing delay of 6 ns. The timing can be met, if the clock period becomes 6 ns, otherwise there will be timing violation in path 1. But as the clock period will increase, it may degrade the pipeline performance. This problem is resolved with the replacement of flip-flop 1 by a latch, because the edge-triggered flip-flop changes the state an edge transition, whereas latch changes state as long as the clock pin is enabled. The latch opens at the same time as flip-flop i.e. 0ns, but it closes at 2.5 ns (negative edge of the clock). So, path 1 has an extra 2.5 ns to borrow from the next cycle. The time borrowed by path 1 = 6ns-5ns = 1 ns, whereas it can use 2.5

ns, so it has a positive slack of 1.5 ns. In such a scenario, there will be no time violation by path 1. Path 2 will start immediately after path 1. Path 2 will add a 1ns delay.

Path 2 could use up to 4 ns (2.5 ns is half period of clock cycle + 1.5 ns positive slack), but it uses only 1 ns. The data capturing of flip-flop 2 is available at 4 ns. The flip-flop's rising edge occurs at 5 ns. So, a positive slack is of 1 ns.

- **Ques: 24. What do you understand by time stealing?**

Solution:

Time stealing is the concept of adjusting the clock phase at flip-flop2 so that data arrival time at the capture edge of flip-flop2 will not violate the timing constraints. Time stealing is used when specific logic partition needs additional time which should be deterministic at the start time.

- **Ques: 25. What are the main characteristics of the time borrowing concept?**

Solution:

1. Time borrowing can be multistage.
2. In time borrowing, both data launching and capturing should be completed using the same phase of the same clock. If the launching and capturing are out of phase, time borrowing will be deactivated.
3. It should be held in the same clock cycle.
4. Time borrowing slowdowns the data arrival time.
5. Time borrowing affects setup slack calculation.
6. Hold slack calculation is not affected with time borrowing because the fastest data is used by hold time.

- **Ques: 26. What is the difference between time borrowing and time-stealing?**

Solution:

1. Time borrowing applies to latch-based design whereas the time stealing applies to flip-flop-based design.
2. The method of borrowing time from the shorter paths of the next design stages to the bigger path is known as time borrowing. The adjustment of the clock cycle for flip-flop2 as per the data arrival time is known as time stealing.
3. In the time borrowing concept, there is no interference with the clock phase. The latch uses the previous cycle slack automatically through the

pipeline. Whereas time stealing, steals the time from the next stage and it results in less time to the next stage. The designer will take care that the next stage delay should be lesser than the difference between the clock period and phase shift.

- **Ques: 27. How will you calculate negative borrow time and maximum borrow time?**

Solution:

The negative time borrow can be calculated as the difference between data arrival time and clock edge.

$$Negative\ borrow\ time = Arrival\ time\text{-}clock\ edge$$

The negative borrow time states that there is no borrowing takes place.

$$Maximum\ borrow\ time = clock\ pulse\ width\text{-}library\ setup\ time$$

Along with library time at end of latch, clock latency is also subtracted from the clock pulse width to achieve the maximum borrow time.

- **Ques: 28. What do you mean by positive, negative and zero slack?**

Solution:

The slack is the factor that determines the speed or frequency of the specific design. It is related to the timing path and can be calculated as:

$$Slack = Actual\ time\text{-}desired\ time$$

The negative slack means, there is some timing violation. The design has not achieved a specific speed or frequency.

The positive slack means, the design is achieving the specific speed or frequency. It has some extra margin as well.

The zero slack signifies that there is no margin, but the designer is already working on the exact speed or frequency.

- **Ques: 29. How will you measure slack for setup and hold time?**

Solution:

The slack for setup time as well as hold time can be calculated as:

$$Setup\ slack = Data\ required\ time\text{-}data\ arrival\ time$$
$$Hold\ slack = data\ arrival\ time\text{-}data\ required\ time$$

So, the difference between the actual and desired time of data to clock time is referred to as slack time. The arrival time refers to the time by data to travel through the timing path. Th time taken by the clock to transverse through the clock path is known as the required time.

- **Ques: 30. Enlist the ideal conditions for the timing path?**

Solution:

The basic static timing equations are as follow:

$$Clock\ period > Tcq + Tpd + Tsu \tag{1.1}$$

Where

Tcq is the maximum time from clock to output; Tpd is maximum propagation delay time through the logic and Tsu is maximum setup time.

$$Hold\ time < Tmin\ (R) + Tmin\ (logic) \tag{1.2}$$

Where

Tmin (R) and Tmin (logic) are the minimum delays by register and logic respectively.

$$Clock\ period + clock\ skew > Tcq + Tpd + Tmin\ (logic) \tag{1.3}$$

Where clock skew is a spatial delay of the clock.

$$Hold\ time + clcok\ skew < Tcq + Tpd + Tmin\ (logic) \tag{1.4}$$

$$Clock\ period - jitter\ (worst\ case) > Tcq + Tpd + Tsu \tag{1.5}$$

$$Hold\ time + jitter\ (worst\ case) < Tmin(R) + Tmin(logic) \tag{1.6}$$

The worst case of jitter is the situation where the rising edge is late and falling edge is early.

$$Worst\ case\ jitter = 2 \times jitter \tag{1.7}$$

So, the maximum frequency of operation (1/clock period) is dependent on the maximum Tcq, Tpd, and Tsu. The setup violation can be fixed by varying clock frequency and temperature. The temperature will further reduce the threshold voltage and makes the device faster. The hold time violation cannot be fixed by changing the clock frequency.

- **Ques: 31. What do you mean by clock skew? What is positive, negative and zero clock skew?**

Solution:

In synchronous circuits, if the clock signal arrives at different components at different times, although the clock signal is generated from the same source. The reasons for clock skew may be temperature variation, capacitor decoupling, wire interconnect length or material imperfection.

When the transmitting source receives the clock tick before the receiver, it is known as the positive clock skew. The positive clock skew enhances the operating frequency and makes the hold time tougher.

When the transmitting source receives the clock tick after the receiver receives it, this is known as the negative clock skew. The negative skew decreases operating frequency.

When there is synchronization between transmitter and receiver for clock arrival, it is known as zero cock skew.

Clock skew is also known as clock uncertainty.

- **Ques: 32. How does the clock skew violate setup and hold time constraints?**

Solution:

The clock skew can cause two types of violation:

(a) Setup violation
(b) Hold time violation

When the clock signal travels at a slower speed than required, then the integrity and synchronization between source and destination are destroyed. The previous data is not stored for sufficient time to be clocked through properly, it is known as hold time violation.

When the clock signal travels faster, the destination receives the clock tick before the source, it causes set up the violation. As the data reaches late as well as it is not stable before the clock signal, it leads to setup violation.

- **Ques: 33. What do you mean by clock jitter?**

Solution:

When the clock edge deviates from its ideal position, it is known as clock jitter. The reason for clock jitter may be noise, power supply variation or interference due to neighborhood circuits. The clock jitter is shown in Figure 1.6.

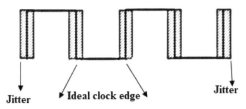

Figure 1.6 Clock jitter.

The jitter may affect the clock signal to be slower or faster that will further violate the setup or hold time constraints. This will degrade the performance or functionality of the chip or circuit. So, it is an important parameter while designing the circuit and timing analysis.

- **Ques: 34. How many types of clock jitter are there?**

Solution:

There are four types of clock jitter, which may be present in the circuitry.

1. Period jitter

The average value of clock period is the deviation over the RMS value of the deviation of 10,000 clock cycles. It is also known as the peak-to-peak period jitter.

2. Cycle to cycle jitter

Within a random 1000 clock cycles, the deviation between two adjacent clock cycle edges is known as cycle to cycle jitter. It measures the difference between minimum clock edge change to the maximum clock edge change.

3. Phase jitter

This is rapid and short-term fluctuations due to phase noise in the frequency domain. It can be translated into jitter values.

$Phase\ noise = \frac{signal\ power}{noise\ power}$, normalized at 1Hz bandwidth at a given offset from the carrier signal.

4. Time interval error (TIE) jitter:

It determines how far each active edge varies from the corresponding edge of the ideal clock. The RMS TIE measures the standard deviation of timing error.

- **Ques: 35. Which type of jitters can be used to determine high-frequency jitter?**

Solution:

Cycle to cycle jitter is used for determining high-frequency jitter. In the random group of clock cycles, it represents the peak value of the clock jitter.

- **Ques: 36. What do you mean by reset? How many types of resets are available?**

Solution:

The reset is the parameter to make the circuit initialize. As hardware has not self-initialization property, so, reset forces it to a known state. During the simulation, reset takes the circuit to the starting and in real hardware, the reset powers up the circuit. There are two types of reset:

1. Synchronous reset
2. Asynchronous reset

- **Ques: 37. Explain the concept of synchronous reset along with its advantages and disadvantages?**

Solution:

The synchronous reset means it is sampled with the clock. The synchronous reset will not be activated until the clock edge is high. The reset should be stretched so that it is visible during the clock signal.

Advantages:

1. A complete synchronous circuit is achieved.
2. The problem of clock glitches is reduced.
3. Deassertion will happen within 1 clock, so it will meet the reset recovery time constraints.

Disadvantages:

1. It is not suited for clock gated circuits.
2. It makes the process slow.
3. A clock signal must be present always.
4. The reset signal should be wide enough to be visible through the clock signal.
5. Reset signals may interfere with other signals during timing analysis and synthesis.

The synchronous reset will be used when a designer needs a complete synchronous circuit, which has no metastability or clock glitch issues.

- **Ques: 38. Explain the concept of asynchronous reset along with its advantages and disadvantages?**

Solution:

An asynchronous reset will be activated as soon as the reset signal is high/enabled. It is not dependent on the clock signal. There is no need to wait for the clock signal.

Advantages:

1. No need to wait/activate the clock signal.
2. It makes the process faster.
3. Reset has the highest priority.

Disadvantages:

1. Metastability may occur.
2. Chances of clock glitches may occur.

The asynchronous reset will be used when the chip needs to be powered up before the clock signal.

- **Ques: 39. What do you mean by reset assertion and reset De-assertion?**

Solution:

Reset Assertion: Activate/apply the reset i.e. when the reset signal is logically true.

Reset De-assertion: The release/disable of reset i.e. when the reset signal is logically false.

During asynchronous reset, the de-assertion may cause metastability because it may be the case that some flip-flops come out of reset before others.

During an asynchronous reset, the reset assertion and de-assertion should meet the minimum required pulse width.

- **Ques: 40. What is reset recovery time?**

Solution:

Reset recovery time is timing validation rule for clock and reset signal. It is similar to the setup time rule. Reset recovery time is the time between de-assertion of reset and activation of the next clock edge. The recovery timing check ensures the same i.e. as soon as asynchronous reset becomes disable or de-assert, the check should ensure that there will be sufficient time to recover so that the next clock will be effective.

When reset is released, some time is taken to make it stable. So, reset recovery time is the minimum time required between the release of reset and arrival of the next clock edge.

For example, let us take a case of a flip-flop, if the clock edge becomes active immediately after the removal of reset, it will take the flip-flop into an unknown state, which will violate the timing constraints. The Figure 1.7 shows reset recovery time.

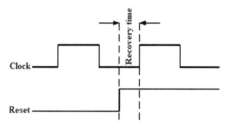

Figure 1.7   Reset recovery time.

• **Ques: 41. What do you mean by reset removal time?**

Solution:

Reset removal time is timing validation rule for clock and reset signal. It is similar to hold time rules. The reset removal time depicts the minimum amount of time between the clock edge and the release of the reset signal.

So, reset removal time is the minimum time required between the arrival of the clock edge and de-assertion of reset.

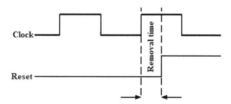

Figure 1.8   Reset removal time.

It should be taken care that the de-asserted reset signal should not get captured on the clock edge at which it is launched. After the clock edge, the reset signal must be stable for some time i.e. removal time. The reset removal time is shown in Figure 1.8.

• **Ques: 42. Explain the concept of a lockup latch?**

Solution:

The latch lockup is the concept in STA where a higher clock skew is present. The lockup latch is just like a transparent latch, which is placed at that point, where clock skew is maximum. To reduce the clock skew and follow the hold time constraints, the lockup latch is used, during design for testability. The Figure 1.9 shows the concept of lockup latch.

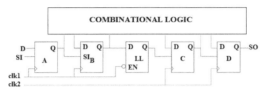

Figure 1.9 Latch lockup.

The clock skew occurs mostly in the systems where multiple clocks are used. The clock skew can occur during shift and capture time. The clock skew can be minimized during a shift by grouping all the flip-flops which are run by the same clock. To remove clock skew completely, the lockup latch should be inserted where the domains cross. This will solve the problem of clock skew during shift.

In scan-chain, the lockup latch will act as an end-point. The scan chain can be reordered, by grouping the cells from starting to lockup latch as one domain, and from lockup, latch to the last cell as a second domain. That way makes the clock grouping preserved.

The timing path will be divided as:

Domain 1: Launch flip-flop to lockup latch

Domain 2: Lockup latch to capture flip-flop

The lockup latch can be placed in between cells automatically or by using a scan chain order file.

There may be multiple clock paths between clock domains that are available during capture. The clock skew during capture can be reduced by a pulse one clock per pattern.

• **Ques: 43. If the clock skew is large, can you use buffers to avoid hold time constraints violation?**

Solution:

It is discouraged practice to use the buffers when clock skew is large. As the number of buffers will be increased, which will automatically degrade the performance of the circuit as area and power factor will also be enhanced. It will increase the chances of on-chip variation (OCV). The optimized solution for handling large clock skew and hold time constraints is the insertion of a lockup latch.

- **Ques: 44. What are the advantages of a lockup latch?**

Solution:

1. It is power and area efficient.
2. The device can handle more OCV (on-chip variation) easily.
3. This is the robust method to deal with hold time constraints during scan shift mode.
4. It prevents data corruption i.e. data overridden which occurs due to clock skew.

- **Ques: 45. Is there any difference between the lockup latch and lockup register?**

Solution:

As the lockup latch occupies approximately half of the area than the lockup register, so lockup latch is having an optimized solution in terms of power and area as compared to the lockup register. During negative lockup latch, there is no need to worry about timing constraints at functional frequency. But this is not for the lockup register. So, the lockup latch is more prevalent than lockup register, during the design process and timing analysis.

- **Ques: 46. What do you understand by the term 'clock latency'?**

Solution:

When there is a difference between the arrival of the clock from source to pin. It is further divided into source latency and network latency. Network latency measures how fast the network is running and source latency specifies the propagation delay from the source of the clock to clock port.

- **Ques: 47. Is the term clock skew and global skew the same?**

Solution:

No, the clock skew and global skew both are different in terms of connections. The global skew is related to skew in between two mutually exclusive flip-flops, i.e. which are not related by fan-in or fan-out. The skew between two independent flip-flops is known as global skew, whereas the skew in between two dependent flip-flops are known as clock skew.

- **Ques: 48. Can you fix the timing path? If yes, then give at least three ways to fix the timing path?**

Solution:

Yes, timing paths can be fixed. It can be done by any of the following ways:

1. Logic optimization
2. Use of macros
3. Placement of logic/capture/launch flip-flop
4. Pipeline can be enhanced
5. Replicate drivers and split number of receiving gates
6. Divide large serial operation into multiple smaller length parallel operations.
7. Switch to the cells having low threshold voltage, high gate leakage, and fast speed.
8. Use one hot encoding register, that will increase the speed of operation.
9. Use power trade-off techniques.
10. Physical design techniques to reduce capacitance and speed up the wire delays.

- **Ques: 49. What is a false path in static timing analysis?**

Solution:

The false path refers to a path that is not required to be optimized during timing analysis. It means it is not necessary to complete the capture and launch a task in the same clock cycle. It is known as a false path. It is not optimized by the timing optimization tool.

- **Ques: 50. What is one hot encoding method?**

Solution:

In the one-hot encoding technique, the number of flip-flops is increased and combinational logic is minimized. It is a state assignment method in a finite state machine. It assigns one flip-flop to each state of FSM. The number of interconnections between logic gates is reduced, which further reduces the propagation delay and speed up the finite state machine.

- **Ques: 51. What is the concept of a multicycle path?**

Solution:

Usually, the data setup and hold operation are done during a single clock pulse. But there are some cases where launch and capture can take more than one clock cycle i.e. combinational delay between launch and capture edge is more than one cycle. This is known as a multicycle and timing path through this combinational logic is known as the multicycle path. Although the data is captured during the same clock cycle, in the case of the multicycle path, the capture edge of flip-flop becomes active after a specific number of cycles. Similarly, the designer will take care of the fact that data will be launched not after every single clock cycle. In such cases, the timing tool will be provided by exception or overridden flag so that it can postpone the launch and capture check after one clock cycle.

- **Ques: 52. How a multicycle path is achieved by the timing tool?**

Solution:

Generally, it was expected to complete the launching of data and capturing the same within one clock cycle. But there may be the scenario, where it needs more than one clock cycle to complete the launch and capture process. It can be accomplished by instruction:

Set_multi_cycle_n -from <start point> -to <end point>

Here 'n' specifies the number of clock cycles needs to complete the launch and capture task. This has been instructed to timing tool to verify and analyze the timing path constraints specifications and violations.

- **Ques: 53. What do you mean by cell delay and net delay?**

Solution:

A wire connecting pins of standard cells is known as the net. The timing delay between input and output pin of a cell is known as cell delay and the timing interconnect delay between the driver pin and load pin is known as a net delay. The stage delay is the sum of net delay and cell delay.

The net delay is the time needed to charge or discharge all the parasitic of the net i.e. resistive, capacitive, inductance, etc.

If the physical wire is not present, we cannot estimate the net delay. Because the accurate value of parasitic depends on the dimensions of the wire.

- **Ques: 54. Enlist the parameters on which net delay or cell delay depends?**

Solution:

The net delay or cell delay depends on the following parameters:

1. Input skew
2. Library setup time

3. Library delay model
4. Cell load characteristics
5. Cell drive characteristics
6. Operating conditions
7. Back annotated delay
8. Wire load model
9. External delay

- **Ques: 55. What is the worst delay and best delay?**

Solution:

Every logic gate and net have min and max delay. In static timing analysis, the maximum delay is known as the worst delay and the minimum delay is known as the best delay. The rise and fall delay are also categorized as min and max delay.

- **Ques: 56. Enlist types of delay models used to estimate the delay?**

Solution:

1. Wire load model
2. Elmore delay model
3. Lumped capacitor model
4. Lumped RC model
5. Distributed RC model
6. RLC model
7. Transmission line model

In design, if a particular delay model is applied, then the same model applies to all cells in a particular library. In a single library, multiple delay models cannot be applied.

- **Ques: 57. What is static sensitization?**

Solution:

A path is the static sensitized path when all the side inputs of the path hold non-controlling values. The controlling (non-controlling) value for the AND gate is 0(1). The static sensitization is sufficient for a path to be a true path in the circuit.

A path is statically co-sensitized if the input corresponding to the path is consistent with the value at the output of each gate on the path.

In a co-sensitized path, if path input is controlling then side inputs can also be sensitizing.

• Ques: 58. What do you mean by signal integrity issues?

Solution:

A set of design issues such as crosstalk, cross-coupling effect, electromigration, and IR drop is called signal integrity issue. A small variation on the single die can violate the design of the whole chip. In an integrated circuit, a wire is routed to another wire using some insulator. An increase in signal value of one wire may vary the signal value of another interconnected wire, in this way, the signal will lose its integrity.

• Ques: 59. What do you mean by crosstalk?

Solution:

Due to the cross-coupling of the capacitor, the signal at one net/wire can interfere with the signal on neighboring net/wire. This disruption of the signal is known as crosstalk. This may further violate set up and hold time violation. The crosstalk creates undesirable voltage spikes known as glitches. There is a possibility of functionality errors due to glitches and timing errors due to deviation in signal timings.

• Ques: 60. How can you avoid crosstalk?

Solution:

1. Increase the spacing
2. Introduce multiple vias
3. Insertion of buffer
4. Shielding
5. Increase the slew rate
6. Use the guard ring

• Ques: 61. How the spacing reduces the crosstalk?

Solution:

When the spacing between the two conductors is more, the width is increased. The cross-coupling will be reduced and consequently, the crosstalk will be reduced.

• Ques: 62. How multiple vias are used to reduce crosstalk?

Solution:

As multiple vias are introduced, the resistances will be in parallel, which will reduce the RC delay and further decrease the crosstalk accordingly.

- **Ques: 63. What is the difference between crosstalk noise and crosstalk delay?**

Solution:

If two signals are close enough, they can cause crosstalk due to coupling capacitance.

Crosstalk delay:

When the one net is switching at a faster rate and the other is switching at a slower speed, due to crosstalk, the speedy net will boost-up the slower net. This is known as crosstalk delay, which is due to timing errors of signals.

Crosstalk noise:

In the case of crosstalk noise, one net is idle (either at logic 1 or logic 0) and the other net is in transition mode (switching from 0 to 1 or vice versa). There may be the introduction of unwanted signal transition due to the coupling capacitor. The reason for crosstalk noise is charge storage effect, power supply or substrate noise. The crosstalk noise analysis tool determines the worst-case glitch on the idle net. The commands for noise analysis are report_noise, check_noise, and update_noise.

- **Ques: 64. Elaborate on the concept of OCV (on-chip variation)?**

Solution:

All devices along a chip should run at a specific speed and interconnects should be either at the worst-case or best-case corner. But due to some variations at the manufacturing level, the speed is not uniform throughout the chip. There is variation in effective channel length and width of transistors. Due to complexities and variations in submicron technologies, the devices with the same size may have different width as compared to the idle condition.

The major on-chip variations are:

1. Variation in the channel length
2. Variation in temperature
3. IR drop variation
4. Variation in transistor width
5. Variation in threshold voltage
6. Variation in interconnects

- **Ques: 65. Enlist any two sources of on-chip variation (OCV)?**

Solution:

1. Etching
2. Photolithography
3. Chemical mechanical planarization

- **Ques: 66. What do you understand by the clock generator and clock distributor?**

Solution:

A clock is a signal that oscillates between low to a high state and vice versa. A network that distributes clock to all clocked elements, for example, buffer and metal network, is known as clock distributor. Whereas, the clock generator is an electronic circuit that produces timing signals. The clock generator is used in the synchronization of a circuit. The basic components of the clock generator are the amplifier and resonant circuit.

- **Ques: 67. What are global chip-to-chip variation and local on-chip-variation?**

Solution:

The performance difference between the die is known as global chip-to-chip variation. It is modeled as operating corners. Within the same die, if there is a performance difference between the transistors, it is known as local on-chip variations. It is modeled as an added derating factor to skew calculations.

   The variation constraints may be:

1. The thickness of the oxide layer
2. Transistor channel dimensions (length and width)
3. The number of doping atoms

- **Ques: 68. What are the two main clock distribution styles in VLSI?**

Solution:

There are two clock distribution systems:

1. Clock tree, also known as Clock Tree Synthesis (CTS). It is placing and routing clock tree elements.
2. Clock mesh or Clock grid distribution system

• **Ques: 69. What is the clock grid distribution system?**

Solution:

The main goal of the clock grid distribution system is to provide uniform delay from the source of the clock to the end receiver i.e. flip-flop, latches, etc. It minimizes the clock skew by providing the concept of stages. The number of stages is dependent on the size of the chip or process technology. It is desirable to have a minimum number of stages. The clock distribution system targets to have the same delay in all stages.

There are two distribution stages; one stage is from the clock source to the boundaries of the block and another stage is a distribution from block boundaries to block. The clock grid distribution system is shown as in Figure 1.10.

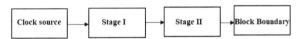

Figure 1.10   Clock grid distribution system.

In stage two, the clock distribution is done inside the block through a fixed number of stages through the mesh/grid of clock buffers. The clock buffers should be symmetric.

• **Ques: 70. Explain the concept of the clock mesh distribution system?**

Solution:

In the clock mesh distribution system, the main clock signal is divided into parallel paths using drivers. The array of buffers is cross-connected in a metallic mesh. The driver feeds these buffers. It routes the path to clock sinks. A resonant structure is created using mesh cross-links. The delay of buffers is terminated because of the resonant structure. The clock mesh distribution system is shown in Figure 1.11.

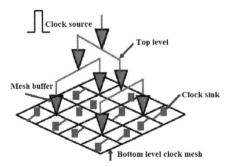

Figure 1.11   Clock mesh distribution system.

This kind of clock distribution system is used in high-speed micropro-cessors. Usually, clock routes are shielded to reduce the coupling effect and variations due to coupling. Due to the variation tolerance nature of the clock mesh system, the problem of clock skew is reduced to a great extent.

- **Ques: 71. What is a clock tree distribution system?**

Solution:

In the clock tree distribution system, the clock is distributed to all receivers i.e. flip-flop, counter, etc. In clock tree synthesis, the number of stages is optimal. It is not necessary to have an equal number of stages throughout the distribution system. As the number of stages is less, so the power dissipation is also low. The clock tree distribution system is shown in Figure 1.12.

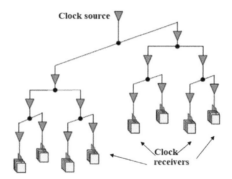

Figure 1.12   Clock tree distribution system.

For slow clock designs, the clock tree distribution is best suited.

- **Ques: 72. What is the difference between the clock mesh and clock tree-type distribution system?**

Solution:

Clock mesh distribution system:

1. Presence of mesh net, which smooths the arrival time difference from multiple mesh drivers.
2. Mesh drivers are connected to mesh net as the multi-driven net.
3. It produces much lower clock skew and clock insertion delay.
4. The design stages are more.
5. Power dissipation is high.
6. Complexity in implementation
7. More routing resources are required to produce clock meshes.

Clock tree distribution system:

1. A mesh net is not required.
2. number of design stages is optimal.
3. Power dissipation is low.
4. The clock tree has a clock source, clock tree cells, clock gating cells, buffers, and load.
5. It is best suited for slow clock circuit designs.

• **Ques: 73. Are clock tree synthesis and clock tree distribution, the same thing?**

Solution:

No, clock tree synthesis and clock tree distribution are not the same things. The clock tree synthesis is used to design the clock tree distribution system. It is used to minimize the clock insertion delay as well as clock skew. For clock tree synthesis, ideal clock arrival times are used, whereas the clock tree distribution system uses real clock arrival time.

• **Ques: 74. What is the need for clock gating?**

Solution:

The clock gating means controlling the clock toggling activity. As the clock drives a lot of elements in a circuit, it consumes a lot of power. The ability to turn off the clock toggling, when not required, is known as clock gating.

In synchronous circuits, clock gating is used to save dynamic power. An extra logic circuit is used which disables the unused clock states. In RTL (register transfer level), the clock gating is commonly used to reduce the size of the die as well as dynamic power consumption. It does not affect the functionality of the design. The clock gating prunes the clock tree i.e. disable the switching of flip-flops. The switching of flip-flop consumes power. By using clock gating, switching power consumption is zero. The clock gating technology saves the area of the die. The concept of clock gating is shown in Figure 1.13.

Figure 1.13 Clock gating.

Clock gating functionally requires only an AND or OR gate. The other input of the AND gate is used to turn off the clock for inactive receivers. Thus, it is an efficient power-saving technique.

- **Ques: 75. What is the significance of CRPR in static timing analysis?**

Solution:

In static timing analysis, CRPR stands for clock reconvergence pessimism removal. The static timing analysis is based on a worst-case analysis. In setup analysis, it uses the slowest possible launch path and fastest capture path. If launch and capture share a common path, the worst case of STA becomes pessimistic as in a common path, fast and slow path cannot happen simultaneously. The CRPR is the accuracy limitation of STA.

- **Ques: 76. What is DEF and what is its use?**

Solution:

The DEF file is a designed exchange format. It is used to describe:

1. Physical aspects of design such as die size, connectivity, macros, etc.
2. Floorplanning information such as standard cells, placement, and routing, etc.
3. The physical representation of power and signal routings, pins, etc.

- **Ques: 77. Explain the term metastability?**

Solution:

In a flip-flop, if the setup and hold time violation takes place, it results in an unpredictable state known as the metastable or quasi-stable state. It can cause a system failure in digital devices such as FPGA, ASIC, etc. In the metastable state, the circuit is unable to settle at either logic 0 or logic 1 within the stipulated time period, this will further fail the system functionality. It happens due to the toggling of flip-flop during the clock transition.

So, when setup and hold time is violated and the output of flip-flop inside the FPGA is unknown or indeterministic, this condition is known as metastability.

- **Ques: 78. What are the effects of metastability?**

Solution:

1. If fan-out is high, the circuit will go to a metastable state, the flip-flop will toggle unintentionally. There will be unexpected behavior of the system.
2. The circuit will draw excessive current.
3. The output will have non-deterministic behavior.
4. The output of the clocked pass gate does not charge properly.
5. The circuit does not meet timing constraints.

• **Ques: 79. What are the reasons for metastability?**

Solution:

1. Slow transition timing constraints at the input and output level (rise time and fall time).
2. Low V_{DD}
3. High parasitic capacitances
4. Cross talk.
5. If the input is an asynchronous signal
6. High clock skew
7. Excessive combinational delay

• **Ques: 80. How metastability can be avoided or tolerated in a circuit?**

Solution:

If input data meet setup and hold time constraints, the problem of metastability can be reduced to an extent. If the signals are generating from different clock domains, it is difficult to control metastability.

1. The clock period should be precise to avoid delay.
2. Add one or more successive synchronizing flip-flops to the synchronizer.
3. Use metastable hardened flip-flops.
4. Provide the needed settling time.
5. Receive each asynchronous signal by clocking it into only one flip-flop.
6. Use asynchronous reset.
7. Use the metastability filter, but it will increase slack.

• **Ques: 81. Can you synchronize between two clock domains?**

Solution:

Yes, two clock domains can be synchronized by using either synchronizer or asynchronous FIFO (if high performance is required). The asynchronous FIFO has two separate interfaces, one clock for reading and another clock is for writing or data extraction purposes.

• **Ques: 82. What is the role of synchronizer?**

Solution:

The synchronizer is used to avoid metastability. It is a digital circuit which is used to convert asynchronous or signals from different clock domains into the receiver's clock domain so that capturing would not cause any metastability issue. It provides sufficient time for the clock signal to settle down the metastable output in the receiver's clock domain.

• **Ques: 83. Design a D latch using 2:1 multiplexer?**

Solution:

The D latch is designed using 2:1 mux. The clock acts as a select line. When the clock is low, the output is fed back to D0 i.e. hold the output state. When the clock is high i.e. logic 1, the data D is transferred to the output terminal.

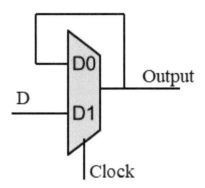

$$Output = \overline{Clock}.D0 + Clock.D1$$

Figure 1.14   D Latch using 2:1 multiplexer.

Thus, in Figure 1.14, the D latch is designed using a 2:1 multiplexer.

• **Ques: 84. How latch and flip-flops are related?**

Solution:

When the two D lathes are connected back to back (as shown in Figure 1.15), it forms a flip-flop.

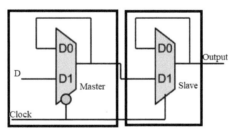

Figure 1.15   Flip-flop using latch.

Here, one latch acts as a master flip-flop, other acts as slave flip-flop. As we know that latch is level sensitive and a flip-flop is an edge-sensitive e device. The first latch is low level and the second latch is high level. It forms a rising edge sensitive D flip-flop. Latch consumes less power than a flip-flop. There are chances of glitches in latch is more than flip-flop.

- **Ques: 85. The device delay is dependent on which factors?**

Solution:

The speed of the device is directly proportional to the following parameters:

1. Width of the device
2. Clock slew rate
3. Load capacitance

- **Ques: 86. What is the difference between statistical and conventional STA?**

Solution:

The application of probability distribution in determining possible circuit outcomes, for variation in the gate and interconnect timings is known as statistical STA. It is different from conventional/deterministic/traditional STA in the following ways:

1. In statistical STA, there is no chance of miss paths, as it does not have any vectors.
2. It can be used for circuit optimization.
3. The run time is linear.
4. It cannot handle spatial correlation within the die, which is possible in the case of deterministic STA.
5. There are correlational problems while using statistical STA, it needs more corners to resolve design issues.

- **Ques: 87. Enlist the major tools that are available for STA?**

Solution:

1. Cadence Encounter
2. Synopsys Primetime
3. Altera Quartus II
4. IBM Eins Timer

- **Ques: 88. Show the RTL design flow?**

Solution:

HDL→RTL Synthesis→Netlist→Logic optimization→Physical design →Layout

- **Ques: 89. What is the front end and back end design?**

Solution:

The frontend and backend design flow are shown as in Figure 1.16:

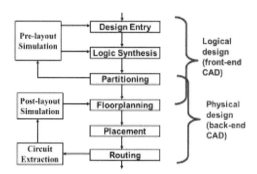

Figure 1.16   Frontend and backend design.

1. Frontend design (By customer)
It is the generation of the gate-level netlist

2. Backend design (By vendor)
Gate level netlist GDSII

3. ASIC fabrication (By foundry)
GDSIIASIC chip

- **Ques: 90. Regarding, CADENCE/Xilinx tools, show the design flow of ASIC design or FPGA prototyping?**

Solution: The Figure 1.17 shows design specification such as IC fabrication and FPGA prototyping and the LVS is shown in Figure 1.18.

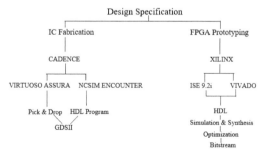

Figure 1.17   Design specification.

LVS (Layout vs. Schematic)

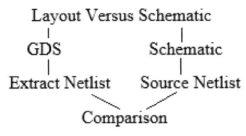

Figure 1.18 Layout Versus Schematic.

PDA (Physical Design Automation): Circuit→Layout

Flow:
Design→Floorplan→STA→Place and Route→Sign-off

Tool	Function
VIRTUOSO	Schematic
SPECTRE	Layout Editor
ASSURA	Layout Verification
QUANTUS	Extraction
TEMPUS	Timing
VOLTUS	Power
PEGASUS	Verification
DFM	Manufacturing

References

1. Gattiker, A., Nassif, S., Dinakar, R., & Long, C. (2001, March). Timing yield estimation from static timing analysis. In *Proceedings of the IEEE 2001. 2nd International Symposium on Quality Electronic Design* (pp. 437–442). IEEE.
2. https://www.vlsisystemdesign.com/crosstalk.php
3. Bhasker, J., & Chadha, R. (2009). *Static timing analysis for nanometer designs: A practical approach.* Springer Science & Business Media.

4. Devgan, A., & Kashyap, C. (2003, November). Block-based static timing analysis with uncertainty. In *ICCAD-2003. International Conference on Computer-Aided Design (IEEE Cat. No. 03CH37486)* (pp. 607–614). IEEE.
5. Malik, S., Martonosi, M., & Li, Y. T. S. (1997, June). Static timing analysis of embedded software. In *Proceedings of the 34th annual Design Automation Conference* (pp. 147–152).
6. Sony, S. (2018). *VLSI interview questions with answers*. Kindle edition.

2

CMOS Design and Layout

2.1 Introduction

A Design Flow in VLSI is the sequence of processes/steps involved in the making of an Integrated Circuit.

Step 1: Logic Synthesis

- RTL conversion into the netlist
- Design partitioning into physical blocks
- Timing margin and timing constraints
- RTL and gate-level netlist verification
- Static timing analysis

Step 2: Floorplanning

- Hierarchical VLSI blocks placement
- Power and clock planning

Step 3: Synthesis

- Timing constraints and optimization
- Static timing analysis
- Update placement
- Update power and clock planning

Step 4: Block Level Layout

- Complete placement and routing of blocks

Step 5: VLSI Level Layout

- VLSI integration of all blocks
- Place and route
- GDSII creation

2.2 CMOS-Design-Flow

The CMOS Design consists of the following steps:

(1) The design of CMOS starts with defining circuit inputs and outputs also called as specifications of the circuit.

(2) Once the detailed list of inputs and outputs is developed from this the design calculations are performed and the circuit schematic for the intended integrated circuit is designed. This developed schematic is then drawn in Computer-Aided Design (CAD) tools e.g. Tanner.

(3) Once the schematic entry is finished then the circuit simulations are carried out and the obtained simulation results are checked with the intended specifications this step is called as pre-layout simulation.

(4) After checking the post-layout simulation results, the next step is the fabrication of the prototype board.

(5) Once the fabricated board comes the testing of the prototype is carried out and the initial specifications are checked, if these results are not matched with the intended specifications then there are two possibilities of error that may be either because of fabrication or initial specification problem. If the prototype board passed all the tests then it is given for mass production. This flow is used for custom IC design.

A custom-designed IC is also called as Application Specific Integrated Circuit (ASIC). Other non-custom methods of designing chips include Field Programmable Gate Arrays (FPGA) and standard cell libraries. The FPGA and standard cell approach is used when low volume and quick design turnaround are important. Most of the chips that are mass-produced such as microprocessors and memories are manufactured using the custom design approach shown in the Figure 2.1.

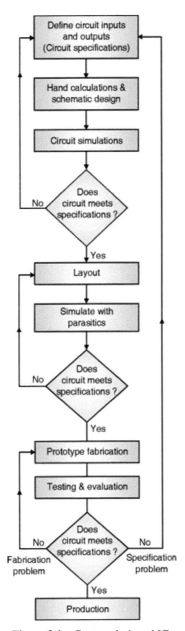

Figure 2.1 Custom designed IC.

2.3 Stick Diagram

VLSI design aims to translate circuit concepts onto silicon. The stick diagrams are a means of capturing topography and layer information using simple diagrams. Stick diagrams convey layer information through color codes (or monochrome encoding). They act as an interface between the symbolic circuit and the actual layout. The stick diagram shows all components/vias and placement of components. The stick diagram is a cartoon of a layout. With the help of a stick diagram, layout planning becomes much easier.

The stick diagram does not show the exact placement of components, the size of transistor and wire lengths/wire widths. Any other low-level details such as parasitic are also not known through the stick diagram. The notation of stick diagram is shown as in Figure 2.2.

2.3.1 Notations of Stick Diagram

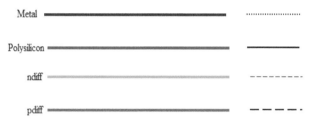

Figure 2.2   Colour scheme of stick diagram.

2.3.2 Rules to draw stick diagram

The rules for stick diagram is depicted in Figures 2.3–2.6.
Rule 1.

When two or more 'sticks' of the same type cross or touch each other that represents electrical contact.

Figure 2.3   Electrical contact.

Rule 2.

When two or more 'sticks' of different types cross or touch each other there is no electrical contact.

Figure 2.4 No electrical contact.

Rule 3.

When a poly crosses diffusion it represents a transistor.

Figure 2.5 Transistor.

Rule 4.

In CMOS a demarcation line is drawn to avoid touching of p-diff with n-diff. All pMOS must lie on one side of the line and all nMOS will have to be on the other side.

Figure 2.6 CMOS using stick diagram.

For Example CMOS inverter, as shown in Figure 2.7

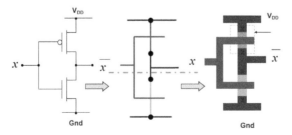

Figure 2.7 CMOS Inverter.

2.4 Design Rules

The term VLSI (Very Large Scale Integration) is the process by which IC's (Integrated Circuits) are made.

There are two basic rules for designing:

1. Lambda Based Design Rule
2. Micron Based Design Rule.

Lambda-based rules:

It allows first-order scaling by linearizing the resolution of the complete wafer implementation. To move a design from 4 microns to 2 microns, simply reduce the value of lambda and so on. However, in general, processes rarely shrink uniformly.

Micron rules:

The micron rule is a normal style for the industry. It can result in as much as a 50% size reduction over lambda rules. The list of minimum feature sizes and spacings for all masks, e.g., 3.25 microns for contact-poly-contact (transistor pitch) and 2.75-micron metal 1 contact-to-contact pitch.

2.4.1 CMOS Lambda 'λ' Design Rules

The MOSIS stands for MOS Implementation Service is the IC fabrication service available to universities for layout, simulation, and test the completed designs. The MOSIS rules are scalable λ rules.

The MOSIS design rules are as follows:

(1) Rules for N-well as shown in Figure 2.8.

1. Minimum width $= 10\lambda$
2. Wells at the same potential with spacing $= 6\lambda$
3. Wells at same potential $= 0\lambda$
4. Wells of a different type, spacing $= 8\lambda$

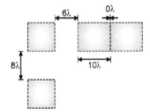

Figure 2.8   Rules for N-well.

(2) Rules for the Active area shown in the Figure 2.9.

    1. Minimum width = 3λ

    2. Minimum spacing = 3λ

    3. Source/Drain active to well edge = 5λ

    4. Substrate/well contact active to well edge = 3λ

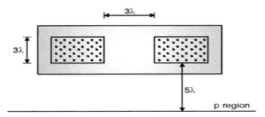

Figure 2.9   Rules for active area.

(3) Rules for poly 1 as shown in the Figure 2.10.

    1. Minimum width = 2λ

    2. Minimum spacing = 2λ

    3. Minimum gate extension of active = 2λ

    4. Minimum field poly to active = 1λ

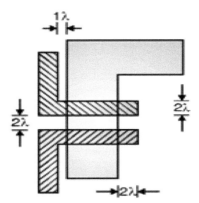

Figure 2.10   Rules for poly 1.

(4) Rules for contact to poly 1 as shown in the Figure 2.11.

    1. Exact contact size = 2 λ × 2 λ

    2. Minimum poly 1 overlap = 1 λ

    3. Minimum contact spacing = 2 λ

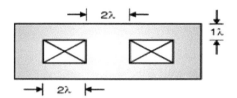

Figure 2.11   Rules gor contact to poly 1.

(5) Rules for contact to active as shown in the Figure 2.12.

1. Exact contact size $= 2\lambda \times 2\lambda$
2. Minimum active overlap $= 1\lambda$
3. Minimum contact spacing $= 2\lambda$
4. Minimum spacing to the gate of transistor $= 2\lambda$

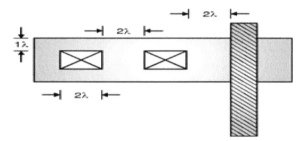

Figure 2.12   Rules for contact to active area.

(6) Rules for metal 1 as shown in the Figure 2.13.

1. Minimum width $= 3\lambda$
2. Minimum spacing $= 3\lambda$
3. A minimum overlap of poly contact $= 1\lambda$
4. A minimum overlap of active contact $= 1\lambda$

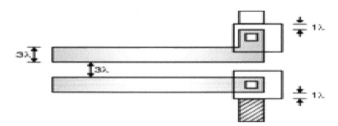

Figure 2.13   Rules for metal.

(7) Rules for via 1 as shown in the Figure 2.14.

    1. Minimum size $= 2\lambda \times \lambda$

    2. Minimum spacing $= 3\lambda$

    3. Minimum overlap by metal 1 $= 1\lambda$

Figure 2.14 Rules for via.

(8) Rules for metal 2 as shown in the Figure 2.15.

    1. Minimum size $= 3\lambda$

    2. Minimum spacing $= 4\lambda$

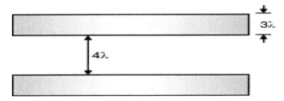

Figure 2.15 Rules for metal 2.

(9) Rules for metal 3 as shown in the Figure 2.16.

    1. Minimum width $= 6\lambda$

    2. Minimum spacing $= 4\lambda$

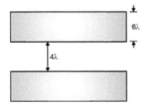

Figure 2.16 Rules for metal 3.

2.4.1.1 Design Rule Check

To ensure that none of the design rules are violated CAD tools named Design Rule Checking (DRC) are used. If DRC is not verified then it leads to the nonfunctional design.

    The layout rules are grouped into three categories that are transistor rules, contact and via rules and well and substrate contact rules.

Transistor rules:

The transistor can be created by overlapping the active and polysilicon layers. The minimum length of transistor equals 0.24 μm which is the minimum width of polysilicon, whereas the width of the transistor is at least 0.3 μm which is the minimum width of the active layer.

The Figure 2.17 below shows the layout of the PMOS transistor.

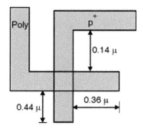

Figure 2.17   Layout of PMOS transistor.

Contact and Via rules:

A contact forms an interconnection between metal and active or polysilicon layer whereas via forms an interconnection between two metal lines. A contact or via is formed by overlapping the two interconnecting layers and provides a contact hole filled with metal between the two.

The Figure 2.18 below shows the contacts and via used in the layout.

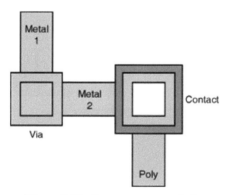

Figure 2.18   Layout contact and via.

Well and substrate contact rules:

For digital circuit design, the well and substrate regions need to be connected to the supply voltages. If this is not done then a resistive path is created between the substrate contact of the transistors and the supply rails which leads to parasitic effects such as latch-up.

2.4.2 Micron-Design-Rules

Micron (μ) Design Rules: Industry uses the micron design rules and code designs in terms of these micron dimensions. The micron design rules are as follows:

(1) Rules for N-well as shown in Figure 2.19.

    1. Width $= 3\mu$
    2. Space $= 9\mu$

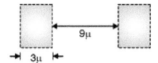

Figure 2.19   Rules for N-well.

(2) Rules for active area as shown in Figure 2.20.

    1. Minimum size $= 3\mu$
    2. Minimum spacing $= 3\mu$
    3. N+ active to N-well $= 7\mu$

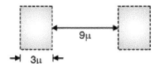

Figure 2.20   Rules for active area.

(3) Rules for poly 1 as shown in Figure 2.21.

    1. Width $= 2\mu$
    2. Spacing $= 3\mu$
    3. Gate overlap of active $= 2\mu$
    4. Field poly 1 to active $= 1\mu$

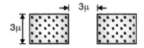

Figure 2.21   Rules for poly 1.

(4) Rules for contact to poly 1 as shown in the Figure 2.22.

    1. Exact contact size $= 2\mu \times 2\mu$

    2. Minimum poly overlap $= 1\mu$

    3. Minimum contact spacing $= 2\mu$

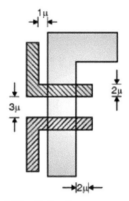

Figure 2.22   Rules for contact to poly 1.

(5) Rules for contact to active as shown in the Figure 2.23.

    1. Exact contact size $= 2\mu \times 2\mu$

    2. Minimum active overlap $= 1\mu$

    3. Minimum contact spacing $= 2\mu$

    4. Minimum spacing to gate $= 2\mu$

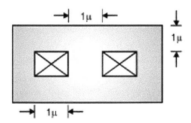

Figure 2.23   Rules for contact to active.

(6) Rules for metal 1 as shown in the Figure 2.24.

    1. Width $= 3\mu$

    2. Spacing $= 3\mu$

    3. Overlap of contact $= 1\mu$

    4. Overlap of via $= 2\mu$

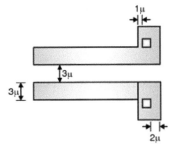

Figure 2.24 Rulles for metal 1.

(7) Rules for metal 2 as shown in the Figure 2.25.

    1. Width $= 3\mu$
    2. Space $= 3\mu$
    3. Metal 2 overlap of via $= 2\mu$

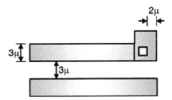

Figure 2.25 Rules for metal 2.

2.5 Layout Design Rules

The layout design rules provide a set of guidelines for constructing the various masks needed in the fabrication of integrated circuits. Design rules are consisting of the minimum width and minimum spacing requirements between objects on the different layers.

The most important parameter used in design rules is the minimum line width. This parameter indicates the mask dimensions of the semiconductor material layers. Layout design rules are used to translate a circuit concept into an actual geometry in silicon.

The design rules are the media between the circuit engineer and the IC fabrication engineer. The Circuit designers require smaller designs with high performance and high circuit density whereas the IC fabrication engineer requires a high yield process.

Minimum line width (MLW) is the minimum MASK dimension that can be safely transferred to the semiconductor material. For the minimum dimension, design rules differ from company to company and from process to process.

To address this issue scalable design rule approach is used. In this approach rules are defined as a function of a single parameter called 'λ'. For an IC process 'λ' is set to a value and the design dimensions are converted in the form of numbers. Typically, a minimum line width of a process is set to 2λ e.g. for a 0.25 μm process technology 'λ' equals 0.125 μm.

2.5.1 Layered Representation of Layout

The layer representation of layout converts the masks used in CMOS into simple layout levels that are easier to visualize by the designers. The CMOS design layouts are based on the following components:

(1) Substrates or Wells: These wells are p-type for NMOS devices and n-type for PMOS devices.
(2) Diffusion regions: At these regions, the transistors are formed and also called an active layer. These are defined by n+ for NMOS and p+ for PMOS transistors.
(3) Polysilicon layers: These are used to form the gate electrodes of the transistors.
(4) Metal interconnects layers: These are used to form the power supply and ground rails as well as input and output rails.
(5) Contact and Via layers: These are used to form the interlayer connections.

For Example, CMOS based 3-input NAND gate, shown in Figure 2.26.

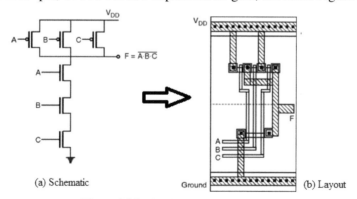

Figure 2.26   CMOS based NAND gate.

Technical Questions with Solutions

- **Ques: 1. What do you mean by design rules?**

Solution:

Design rules are a set of geometrical specifications that dictate the design of the layout

- The layout is a top view of a chip.
- The design process is aided by a stick diagram and layout.
- Stick diagram gives the placement of different components and their connection details, But the dimensions of devices are not mentioned.
- Circuit design with all dimensions is the layout.
- Fabrication process needs different masks, these masks are prepared from layout
- The layout is an Interface between circuit designer and fabrication engineer
- The layout is made using a set of design rules.
- Design rules allow translation of circuit (usually in stick diagram or symbolic form) into actual geometry in silicon wafer
- These rules usually specify the minimum allowable line widths for physical objects on-chip

Example: metal, polysilicon, interconnects, diffusion areas, minimum feature dimensions, and minimum allowable separations between two such features.

- **Ques: 2. What is the need for design rules?**

Solution:

The few reasons are:

- Better area efficiency
- Better yield
- Better reliability
- Increase the probability of fabricating a successful product on Si wafer

If design rules are not followed:

- Functional or non-functional circuit.
- Design consuming larger Si area.
- The device can fail during or after simulation

• **Ques: 3. What is the color coding for different layers?**

Solution:

The various colors representing various layers are as shown in Figure 2.27:

Layer	Color	Representation
N+ Active	Green	▬▬▬▬
P+ Active	Yellow/Brown	▬▬▬▬
PolySi	Red	▬▬▬▬
Metal 1	Blue	▬▬▬▬
Metal 2	Magenta	▬▬▬▬
Contact	Black	X
Buried contact	Brown	X
Via	Black	X
Implant	Dotted yellow	
N-Well	Dotted Green/Black	⌐⁚⁚⁚⌐

Figure 2.27   Color coding.

• **Ques: 4. What do you mean by the stick diagram?**

Solution:

A stick diagram is a symbolic representation of a layout.

- In the stick diagram, each conductive layer is represented by a line of distinct color.
- Width of the line is not important, as stick diagrams just give only wiring and routing information.
- Does show all components/vias, relative placement.
- It does not show exact placement, transistor sizes, wire lengths, wire widths, tub boundaries.

VLSI design aims to translate circuit concepts onto silicon. The stick diagrams are a means of capturing topography and layer information using simple diagrams. Stick diagrams convey layer information through color codes (or monochrome encoding). It acts as an interface between the symbolic circuit and the actual layout.

Stick diagram does not show:

- The exact placement of components
- Transistor sizes

- Wire lengths, wire widths, tub boundaries.
- Any other low-level details such as parasitics

- **Ques: 5. What are the rules of the Stick diagram?**

Solution: The rules of stick diagram are shown in Figures 2.28–2.32.

Rule 1.

When two or more 'sticks' of the same type cross or touch each other that represents electrical contact.

Figure 2.28   Rules for electrical contact.

Rule 2.

When two or more 'sticks' of different types cross or touch each other there is no electrical contact.

Figure 2.29   Rules for non-contact.

Rule 3.

When a poly crosses diffusion it represents a transistor.

Figure 2.30   Rules for transistor.

Red (poly) over Green (Active), gives a FET.

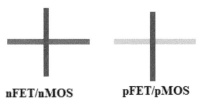

nFET/nMOS   **pFET/pMOS**

Figure 2.31   Rules for FET.

Rule 4.

In CMOS a demarcation line is drawn to avoid touching of p-diff with n-diff. All pMOS must lie on one side of the line and all nMOS will have to be on the other side.

Figure 2.32   Rules for CMOS.

- **Ques: 6. Explain the basic steps to draw stick diagrams?**

Solution:

1. Normally, the first step is to draw two parallel metal (blue) VDD and GND rails.
2. There should be enough space between them for other circuit elements.
3. Draw Demarcation line (Brown) at the center of VDD and GND rails.
4. This line represents the well (n/p-well).
5. Next, Active (Green/yellow) paths must be drawn for required Pull up & Pull-down transistors above & below DL.
6. Draw vertical poly crossing both diffusions (Green & yellow)
7. Remember, Poly (Red) crosses Active (Green/yellow), where the transistor is required.
8. No Diffusion can cross the demarcation line.
9. Only poly and metal can cross the demarcation line
10. N-diffusion and p-diffusion are joined using a metal wire.
11. Place all PMOS above and NMOS below the demarcation line.
12. Connect them using wires (metal).
13. Blue may cross over red or green, without connection.
14. The connection between layers is specified with X.

15. Metal lines on a different layer can cross one another, connections are done using via.

- **Ques: 7. What are the various layout design rules?**

Solution:

There are two types of layout design rules:

1. Industry-standard: Micron rule
2. Lambda based design rules

- **Ques: 8. What do you mean by industry-standard rules?**

Solution:

In industry-standard micron-based rules, all device dimensions are expressed in terms of absolute dimension (μm/nm). These rules will not support proportional scaling

- **Ques: 9. What do you mean by lambda-based design rules?**

Solution:

It is developed by Mead and Conway.

- All device dimensions are expressed in terms of a scalable parameter λ.
- $\lambda = L/2$; L = The minimum feature size of transistor
- $L = 2\,\lambda$

These rules support proportional scaling. They should be applied carefully in the sub-micron CMOS process.

- **Ques: 10. What is the minimum length/width and minimum separation on layers and why?**

Solution:

L is the minimum channel length between drain and source

- The minimum length or width of a feature on a layer is 2 λ, to allow for shape contraction.
- A minimum separation of features on a layer is 2 λ, to ensure adequate continuity of the intervening materials.

- **Ques: 11. What is allowable misalignment for two features on different mask layers?**

Solution:

Two Features on different mask layers can be misaligned by a maximum of 2 λ on the wafer.

If the overlap of these two different mask layers can be catastrophic to the design, they must be separated by at least 2 λ

If the overlap is just undesirable, they must be separated by at least λ.

- **Ques: 12. Enlist the various design rules for CMOS?**

Solution:

Line size and spacing:

1. Metal1:

Minimum width=3 λ, Minimum Spacing=3 λ

2. Metal2:

Minimum width=3 λ, Minimum Spacing=4 λ

3. Poly:

Minimum width= 2 λ, Minimum Spacing=2 λ

4. ndiff/pdiff:

Minimum width= 3 λ, Minimum Spacing=3 λ,

5. Wells:

Minimum width=6 λ,

Minimum n-well/p-well space = 6 λ (They are at the same potential)

Minimum n-well/p-well space = 9 λ (They are at different potential)

6. Transistors:

Min width=3 λ

Min length=2 λ

Min poly overhang=2 λ

7. Contacts (Vias)

Cut size: exactly 2 λ X 2 λ

Cut separation: minimum 2 λ

Overlap: min 1 λ in all directions

- **Ques: 14. What is the contact cut? and what should be its length and breadth?**

Solution:

Metal connects to polySi /diffusion by contact cut. The Figures 2.33 and 2.34 show the contact cut.

Contact area: 2 λ X 2 λ

Metal and poly Silicon or diffusion must overlap this contact area by λ so that the two desired conductors encompass the contact area despite any misalignment between conducting layers and the contact hole.

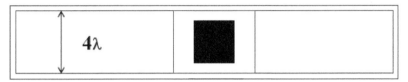

Figure 2.33   Contact cut.

The space between contact cut –contact cut: 2 λ apart, to prevent holes from merging.

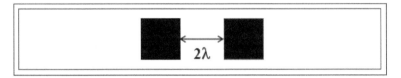

Figure 2.34   Space between contact cut.

- **Ques: 15. What is interlayer contact?**

Solution:

Interconnection between poly and diffusion is done by contacts.

- Metal contact
- Butting contact
- Buried contact

- **Ques: 16. What is the specification for metal contact?**

Solution:

- Contact cut of 2λ* 2λ in oxide layer above poly and diffusion
- The metal used for interconnection
- Individual contact size becomes 4λ* 4λ

- **Ques: 17. What do you mean by 180nm technology?**

Solution:

180nm refers to the smallest available channel length of a CMOS device for that semiconductor process. 180nm is 180 nano-Meters or 0.180 microns.

In VLSI circuit simulation tools such as Cadence is based on 180 nm, 90 nm, 45 nm, etc. With the invention and evolution of transistors, various technologies came into existence and more would continue to come in the future. According to Moore's law, the number of transistors will continue to double every 1.5 years. That means the same silicon area would accommodate several transistors. To achieve this, the transistor size is gradually getting reduced. This we say, transistor size is shifting from one technology node to a smaller technology node by the scaling process. The shifting of technology node helped many leading players in the semiconductor industry like Intel, IBM, AMD, Texas Instruments, etc. to come up with many innovative and highly powerful products day by day. A particular technology gets used by the industries for a while period till the time the next feasible smaller technology node would be ready for implementation. For example, 180 nm technology was used by most of them in the 1999–2000 time-frame, while 90 nm was used in 2004–2005.

45 Nanometer (45 nm) refers to the technology or process used by Intel while producing semiconductor chips processors in 2007–2008.

- **Ques: 18. What is the difference between 180 nm, 90 nm, and 45 nm?**

Solution:

The numbers represent the minimum feature size of the transistor (PMOS or NMOS). The minimum feature size means that during the fabrication process of a transistor, how closely can the transistors be placed on a chip to be used for various purposes. The smaller this size is, the larger the number of transistors can be fabricated on the chip. For example, suppose separate chips are to be designed using 180 nm and 90 nm transistors. Now, the number of 90 nm transistors that can be placed on a particular area of the chip would be more (nearly twice) than the number of 180 nm ones that can be placed on the same silicon area.

The above can also be understood by the fact that the numbers 180 nm, 90 nm, etc. represent the minimum channel length that can be used in fabrication.

Also, these numbers aren't randomly assigned but decided by dividing the previous number by square root of 2 (2 because it is neither too small nor too big). For example, the next technology node after 180 nm was 180 divided by the square root of 2 which comes out to be nearly 130 nm. Likewise, the next after 130 nm will be 130 divided by square root of 2 which is approximately 90 nm and so on. Different technologies are being used today and the transistor size is shrinking day-by-day to lower the cost of production of a chip as smaller the chip, cheaper is to make it.

- **Ques: 19. What is the latest process technology available?**

Solution:

The latest technology node is 7 nm, it was released first by Apple Bionic-A12 for the public in September 2018. It is expected that 5nm technology will be released by the 2020 year.

- **Ques: 20. What is the historical advancement of process technology?**

Solution:

The evolution of the fabrication process provides these improvements, also known as "scaling" trends, without requiring any new circuit/architectural innovation. Major semiconductor companies make elaborate roadmaps to improve IC performance by exploiting process advancement as well as design innovation. Most notable of these companies is Intel, which came up with a "tick-tock" model, where a "tick" means area and power improvements through shrinking of the process technology and a "tock" introduces improvement through a new architecture.

Table 2.1 Historical Advancement of process technology

Technology	10 μm	6 μm	3 μm	1.5 μm	1 μm	800 nm	600 nm	350 nm	250 nm	180 nm
Year	1971	1974	1977	1982	1985	1989	1994	1995	1997	1999

Technology	130 nm	90 nm	65 nm	45 nm	32 nm	22 nm	14 nm	10 nm	7 nm	5 nm
Year	2001	2004	2006	2007	2010	2012	2014	2017	2018	2020

- **Ques: 21. What is the benefit of scaling in VLSI?**

Solution:

The International Technology Roadmap for Semiconductors (ITRS) and since then The ITRS has laid out the foundations of the era of "scaling".

If the size decreases, the effective capacitance decreases, delay decreases and which automatically increases the speed.

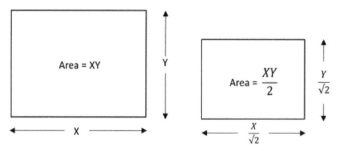

Figure 2.35   Scaling in VLSI.

As shown in the Figure 2.35, a 30% reduction in the dimensions of the rectangle reduces the area by 50%. This means a scaling factor 0.7x (=0.7) is needed to reduce the area to half.

Reduction in the size of the devices reduces effective capacitance, which in turn reduces device delay by 30% (0.7x), making devices run faster. One measure of the fastness of the devices in a process node is operating frequency, which is defined by the frequency of ring oscillator as shown in the Figure 2.36:

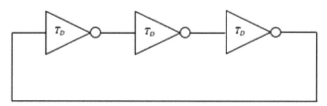

Figure 2.36   Operating frequency.

If the delay of each inverter of the ring oscillator is, the topology of the circuit ensures that the input of the first transistor gets fed back after 3, making the frequency of this oscillator. So, a 30% reduction in delay (0.7x) will correspond to a 40% increase in operating frequency – clear advantage of scaling!

$$0.7 = \frac{1}{\sqrt{2}}$$

Finally, to keep the electric field constant, the voltage in a process is reduced by 30%. This way, each new technology generation doubles transistor density while keeping power consumption the same.

• **Ques: 22. What do you mean by feature size in VLSI?**

Solution:

This scaling factor of 0.7x provides intuition for process nodes roadmap from 180nm to $180 * 0.7 = 130$ nm to $130 * 0.7 = 90$ nm to $90 * 0.7 = 65$ nm to 45 nm, 32 nm, 22 nm, 16 nm and so on. These numbers represent feature size, as shown in Figure 2.37.

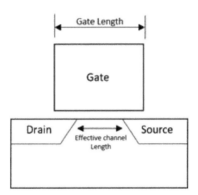

Figure 2.37   Effective channel length.

Before the 32 nm node, the process node roughly corresponded to the minimum value of drawn gate length. And it could be said that the feature size is an indication of the gate length. It is important to highlight here that the actual channel length or effective channel length of the implemented transistor would have been lower than node value considering overlap from source and drain regions on the gate area and thereby reducing channel length. The effective channel length remained constant from 90 nm to 32 nm.

The feature size of any semiconductor technology is defined as the minimum length of the MOS transistor channel between the drain and the source.

Smaller the node $->$ denser the devices $->$ more computing logic in less area $->$ more functionality $->$ faster and lower power per function

• **Ques: 23. Explain the Pitch and node?**

Solution:

International Technology Roadmap for Semiconductors uses 'node' to refer to the smallest feature size on Logic Chips. This is the length of the Gate of MOSFET. Whereas in Memory Chips 'half-pitch' is used to define the smallest feature size.

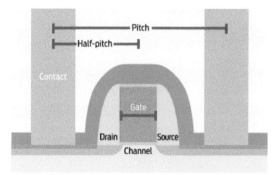

Figure 2.38 Pitch in MOSFET.

Half-pitch is the half distance between two adjacent aluminum pathways. Nodes and half-pitch are illustrated as in Figure 2.39:

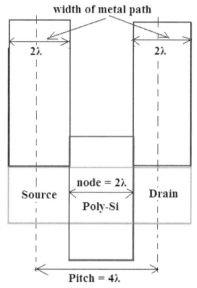

Figure 2.39 Pitch and node.

DRAM half-pitch: The common measure of the technology generation of a chip. It is half the distance between cells in a dynamic RAM chip.

The 32 nanometers (32 nm) node is the step following the 45-nanometer process in CMOS semiconductor device fabrication. "32 nanometer" refers to the average half-pitch (i.e., half the distance between identical features) of a memory cell at this technology pitch in MOSFET is shown in Figure 2.38 level.

- **Ques: 24. What are DC and AC analysis?**

Solution:

While designing an analog circuit, first check BIAS conditions, i.e., DC Analysis. This tells what would happen if you simply turned the circuit on and applied no signal to it. AC analysis to figure out the frequency response of the circuit. The transient analysis will determine how your circuit will behave under non-well-behaved signals. It could be a huge peak load, a peak change in the input, etc. This can be analyzed using Cadence Virtuoso tools.

(a) DC operating point analysis
when DC voltage is applied to the circuit calculates its behavior.

(b) DC sweep analysis
It calculates circuit bias point over a range of values

(c) Transient analysis
It is the circuit response as a function of time. The offset voltage, frequency, and frequency are calculated using transient analysis when a small signal is applied.

(d) AC analysis
The small-signal response of circuit such as gain versus frequency or phase versus frequency.

(e) Noise analysis
Noise contribution from each 'R' and 'semiconductor device' at specified output value.

(f) Noise figure analysis
It tells how noisy a device is.

From the dc analysis of a network, we can determine the node voltage, mesh current and branch voltage and branch current.

From the ac analysis, we can determine the resonance condition, Phase angle, Q-factor, dissipation factor, maximum & minimum impedance.

But using transient analysis, we can determine the charging and discharging time of capacitor and inductor, the transient behavior of series & parallel circuit & steady-state error.

• **Ques: 25. What are the various CADENCE tools and their function?**

Solution:

S.No	Tool	Purpose
1	COMPOSER	Schematic
2	VIRTUOSO	Layout editor
3	SPECTRE	Simulation
4	DIVA	DRC/LVS/Extraction

• **Ques: 26. Explain the various steps used for analog IC design flow?**

Solution:

Design flow:

Design Specification→Schematic Capture→Create Symbol→Simulation →Layout→DRC→LVS→Post Layout Simulation→Foundry

Explanation:

1. Design Specification schematic capture

   • It describes the gate-level design via the schematic editor.
   • Place and connect individual components.
   • Describe the electrical properties of components.

2. Create a symbol

   • If in design, small modules are there, then assign each such module an icon/symbol.

3. Simulation

   • It verifies electrical performance and circuit functionality.

4. Layout

   • It describes the geometry and positioning of each mask layer

5. DRC (Design Rule Check)

   • Created mask layout should confirm the complex set of design rules.

6. Extraction

   • It identifies the device and generates the netlist associated with the layout.

7. LVS

 • Layout versus Schematic compares the original network with extracted mask layout and proves the equivalency.

8. Post Layout Simulation

 • It simulates the extracted cell view, sweep variable, transient/DC analysis.

9. Foundry

 • For chip fabrication.

References

1. Weste, N. H., & Eshraghian, K. (1985). Principles of CMOS VLSI design: a systems perspective. *NASA STI/Recon Technical Report A*, *85*.
2. Hilleringmann, U., & Goser, K. (1995). Optoelectronic system integration on silicon: waveguides, photodetectors, and VLSI CMOS circuits on one chip. *IEEE transactions on Electron Devices*, *42*(5), 841–846.
3. https://www.vlsifacts.com/180-nm-90-nm-45-nm-whats-difference/
4. http://www.electronics-tutorial.net/Digital-CMOS-Design/CMOS-Layout-Design/CMOS-Design-Flow/
5. Wolf, W., Newkirk, J., Mathews, R., & Dutton, R. (1983). Dumbo, a schematic-to-layout compiler. In *Third Caltech Conference on Very Large Scale Integration* (pp. 379–393). Springer, Berlin, Heidelberg.

3

Physical Design Automation

3.1 Introduction

Physical design is one of the steps in the VLSI design cycle. In this step, each component of a circuit is converted into a set of geometric patterns that achieves the functionality of the component. The physical design step can further be divided into several sub-steps. All the sub-steps of the physical design step are interrelated. Efficient and effective algorithms are required to solve different problems in each of the sub-steps. Good solutions at each step are required since a poor solution at an earlier stage prevents a good solution at a later stage.

Physical design is the process of placement and routing (P&R) of ICs (cell-based ASICs, custom ASICS, FPGA). The main concern is cell-based ASICs P&R flow where each cell is assigned a geometric location and connected to other cells by way of metal lines. This is done automatically by EDA tools such as Nitro-SoC (Mentor Graphics), the resulting layouts are almost always correct by construction and design productivity is much better than for manual layout.

3.2 Types of Cell for Physical Design Automation

There are different types of cells used to meet the physical design requirements.

3.2.1 Well Tap Cells

These library cells connect the power and ground connections to the substrate and n-wells respectively. By placing well taps at regular intervals

71

Table 3.1   Mentor graphic tool nitro input-output data

Input/Output Data	File Content	Extension
Design Synthesized Netlist	A gate-level description (physical entity) generated by a synthesis tool from the RTL description when an ASIC library is selected.	.v
Physical Libraries (Lef of GDS file for all design elements like macro, std Cell, IO pads, etc.)	It contains complete layout information and Abstract model for placement and routing like pin accessibility, blockages, etc.	.lef
Timing, Logical and Power Libraries	Contains Timing and Power info	.lib
Constraints	Contain all design related constraints like Area, power, timing	.sdc/.tcl
Floorplanning	Contain floorplanning information if this step is done with a third party tool and needs to be imported	.def/.pdef
QoR report	Transcript a report containing the QoR metrics values: timing, congestion, wire-lengths, area, utilization, power	.log
Run transcript	Report on the details of different operation run by the tool.	.log

throughout the design, the n-well potential is held constant for proper electrical functioning.

3.2.2 End Cap Cells

These library cells do not have signal connectivity. They connect only to power and ground once power rails are created in the design. They also ensure that gaps do not occur between the well and implant layers. This prevents DRC violations by satisfying well tie-off requirements for the core rows.

3.2.3 Decap Cells

The decap cells are temporary capacitors added in the design between power and ground rails to counter functional failures due to dynamic IR drop. Due to this simultaneous switching, a high current is drawn from the power grid for a small duration. If the power source is far away from a flop, the chances are that, this flop can go into a metastable state due to IR drop. To overcome this problem, decaps are added. At an active edge of the clock, when the current requirement is high, these decaps discharge and provide a boost to the power grid. Decaps are placed as fillers. The closer they are to the flop's sequential elements, the better it is.

3.2.4 Spare Cells

Spare cells need to add while the initial implementation. There are two ways to do this. The designer adds separate modules with the required cells. The placement and routing start with spare cells included and must make sure that the tool has not optimized them away. The inputs are tied to power or ground nets, as floating gates should not be allowed in the layout. The outputs are left unconnected. Spare cells can also be added to the design by including cells in Netlist itself.

3.2.5 Filler Cells

Filler cells are used to establish the continuity of the N-well and the implant layers on the standard cell rows, some of the small cells also do not have the bulk connection (substrate connection) because of their small size (thin cells). In those cases, the abutment of cells through inserting filler cells can connect those substrates of small cells to the power/ground nets, that is, those thin cells can use the bulk connection of the other cells.

Technical Questions with Solutions

- **Ques: 1. What do you understand by Physical design? Show all of its stages.**

Solution:

It is the process of transforming a circuit description into the physical layout, which describes the position of cells and routes for the interconnection between them. The physical design process stages are listed below, as in Figure 3.1.

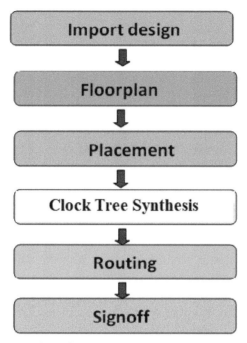

Figure 3.1   Steps for physical design.

- **Ques: 2. Explain all stages of Physical Design Briefly?**

Solution:

The physical design consists of mainly six stages, as shown in Figures 3.2–3.4:

1. Import design:

It is the first stage in physical design. In the synthesis process, the RTL code is converted into a netlist. In this import design stage, all the input files are read by the tool. By using this information, the design process will start.

2. Floorplan:

The floorplan is the process of determining the macro placement, power grid generation and i/o placement. It is the process of placing blocks/macros in the chip/core area thereby determining routing areas between them. It determines the size of the die and creates wire tracks for the placement of standard cells. It creates power straps and specifies a power grid connection. It also determines the I/O, pin/pad placement information.

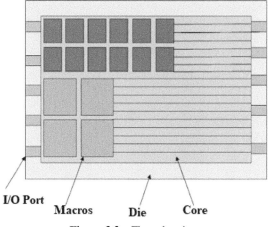

Figure 3.2 Floorplanning.

3. Placement:

Placement is the process of automatically assigning the correct position to standard cells on the chip with no overlapping.

By global placement outside of standard cells will be placed inside roughly.

By the detailed placement, the standard cells will place in site rows (legalize placement).

In the placement stage we check the congestion value by the GRC map.

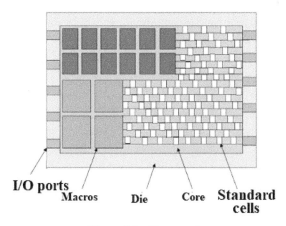

Figure 3.3 Placement.

4. Clock Tree Synthesis:

In this stage, we built the clock tree by using inverters and buffers. In the chip clock signal is essential to the flip flops, to give the clock signal from clock source we built the clock tree. It is the process of balancing the clock skew and minimizing insertion delay to meet timing and power.

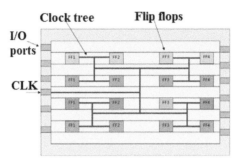

Figure 3.4   Clock tree synthesis.

5. Routing:

Before the routing stage, the connection between the macros, standard cells, clock, i/o port is logical connections. In this stage, we connect all the cells physically with the metal straps.

Routing is divided into two parts

(1) global routing
(2) detailed routing

The global routing will tell for which signal which metal layer is used. Before the detailed routing, all are the logical connections. In detailed routing, the physical connections are done.

6. Signoff:

After the routing, the physical layout of the chip is completed. In the signoff stage, all the tests are done to check the quality and performance of the layout before tape out.

- **Ques: 3. Which design is more complicated than 10MHZ or 100MHz?**

Solution:

100 mhz is a more complicated design because high frequency means a low time period. So, it is difficult to handle the violations in the low time period.

- **Ques: 4. If you have both IR drops and congestion how will you fix it?**

Solution:

(a) Spread macros
(b) Spread standard cells
(c) Increase strap width
(d) Increase number of straps
(e) Use proper blockage.

- **Ques: 5. What are the Tie-high and Tie-low cells?**

Solution:

These are used to connect the gate of the transistor to either power or ground. It avoids direct connection between power and gate of the transistor.

Tie-high:- One terminal is connected to Vdd and another terminal is connected to the gate of the transistor.

Tie-low:- One terminal is connected to Vss and another terminal is connected to the gate of the transistor.

- **Ques: 6. What are the checks to be done before clock tree synthesis?**

Solution:

(a) Placement-completed
(b) Power ground nets-pre-routed
(c) Estimated congestion-acceptable
(d) Estimated timing-acceptable
(e) Estimated max transition/capacitance-no violations
(f) High fan-out nets.

- **Ques:7. What are the power gating cells?**

Solution:

The power gating is to avoid static power dissipation. The power gating cells are

(a) Power switches
(b) Level sifters
(c) Retention registers
(d) Isolation cells
(e) Power controller

- **Ques: 8. What is HFNS (high fan-out net synthesis) and where it is used?**

Solution:

HFNS is the process of buffering the high fan-out nets to balance the load. Generally, at the placement stage, HFNS is performed.it is also performed at the synthesis step using the design compiler.

- **Ques: 9. Mention the checklist after CTS?**

Solution:

(a) Skew report
(b) Clock tree report
(c) Timing reports for setup and hold
(d) Power and area report

- **Ques: 10. Which metal layer will be used for the clock in 7 metal layer design and why?**

Solution:

Metal 4 and 5 are used because the clock nets will consume 30 to 40% of power in the design. So, to reduce the IR drop we are using low resistance metal. Top 6,7 metal layers for power connection and 5,4 for clock nets.

- **Ques: 11. What is LVS (layout vs schematic)?**

Solution:

It is a class of EDA software that determines whether a particular IC layout corresponds to the original schematic of design.

- **Ques: 12. What is shielding?**

Solution:

Placing ground net is placed in between aggressor and victim nets then voltage discharge on the ground net. This will reduce cross-talk.

- **Ques: 13. What is the isolation cell?**

Solution:

These are special cells required at the interface between blocks that are shut-down and always on. It is necessary to isolate the floating inputs.

- **Ques: 14. What is retention flop?**

Solution:

These cells are special flops with multiple power supply. When design blocks are switched off for sleep mode data in all flip flops contained desires to retain state for this retention flops must be used.

- **Ques: 15. What are the i/p required for CTS?**

Solution:

(a) Detailed placement database
(b) Target for latency and skew if specified
(c) Buffers or inverters to build the clock tree.
(d) NDR rules
(e) Clock tree DRC's

- **Ques: 16. What are the CTS goals?**

Solution:

(a) Minimize clock skew
(b) Minimize the insertion delay
(c) Minimize power dissipation

- **Ques: 17. What are the effects of CTS?**

Solution:

(a) Clock buffers are added
(b) Congestion may increase
(c) Non-clock cells may have been moved to the less ideal location.
(d) It can introduce timing and max transition/capacitance violations.

- **Ques: 18. Why HFNS (high fanout net synthesis)?**

Solution:

To balance the load, HFNS performed too many loads will affect the delay numbers and transition time. Because the load is directly proportional to load, by buffering the HFNS the load can be balanced.

- **Ques: 19. What is a hard macro?**

Solution:

The circuit is fixed and we don't know which type of gates using inside. We know only timing information, not functional information.

- **Ques: 20. What is a soft macro?**

Solution:

The circuit is not fixed and we know which type of gates using inside. We know timing information and also functional information.

- **Ques: 21. What is the formula for distance between macros?**

Solution:

Distance between macros = No.of pins * pitch/total layers.

- **Ques: 22. What is CTO (Clock Tree Optimization)?**

Solution:

It improves the clock skew and clock insertion delay by applying additional optimization. CTO is performed during the clock_opt process.

- **Ques: 23. What is the difference between the normal buffer and clock buffer?**

Solution:

Clock buffer having equal rise and fall time but normal buffer not like that. Clock buffers are usually designed such that an i/p signal with a 50% duty cycle produces an o/p with a 50% duty cycle.

- **Ques: 24. Why should we solve setup violations before CTS and hold violations after CTS?**

Solution:

Setup violations depend on data path while hold violations depend on the clock path. Before the CTS clock path is taken as ideal because we don't have skew and transition numbers of the clock path but this information is sufficient to perform setup analysis. The clock is propagated after CTS that's why hold violations are fixed after CTS.

- **Ques: 25. What is global routing?**

Solution:

It is done to provide instructions to the detailed router about route every net. It provides the channels for interconnect to be routed.

• **Ques: 26. What is detailed routing?**

Solution:

It is where we specify the exact location of the wires/interconnects in channels specified by the global routing. Metal layer information of interconnects is also specified here.

• **Ques: 27. What is the use of a virtual clock?**

Solution:

It will help to reduce the time delay of the overall operation. It is logically not connected to any pin of design and physically doesn't exist.

• **Ques: 28. What is MMMC (multi-mode multi-corner)?**

Solution:

It is a combination of mode and corner that is required for a particular timing check such as setup and hold.

• **Ques: 29. What is the difference between hierarchical design and flat design?**

Solution:

The hierarchical design has blocks and sub-blocks in a hierarchy. Flat design has no sub-blocks and it has only leaf cells. The hierarchical design takes more run time and flat design takes less run time.

• **Ques: 30. During power analysis, if you are facing an IR drop problem then how did you avoid?**

Solution:

(a) Increase the power metal layer width
(b) Go for the high metal layer
(c) Spread macros or standard cells
(d) Provide more straps

• **Ques: 31. What are the types of routing?**

Solution:

(a) Global routing
(b) Track assignment
(c) Detailed routing
(d) Search and repair

- **Ques: 32. What are the benefits of SOI technology?**

Solution:

(a) Low parasitic capacitance
(b) High-speed performance
(c) Reduce short channel effects
(d) No latch-up
(e) Low threshold

- **Ques: 33. What are the guidelines for macro placement?**

Solution:

Fly-lines, port communication, macros are placed at boundaries, the spacing between macros, macro grouping, macro alignment, notches avoiding, orientation, blockages, avoid crisscross placement of macros.

- **Ques: 34. What are the sanity checks in PD?**

Solution:

(a) Check_library
(b) Check_timing
(c) Check_design
(d) Report_constraint
(e) Report_timing
(f) Report_QOR

- **Ques: 35. What is the difference between Halo and Blockage?**

Solution:

Halo:- It is the region around the boundary of fixed macros in design in which no other macros or standard cells can be placed. If macros move halo will also move

Blockage:- It can be specified for any part of the design. If we move the block blockage will not move.

- **Ques: 36. Why we apply NDR rules before routing?**

Solution:

Some times with default routing it is very hard to avoid cross-talk, electromigration. Fixing the cross-talk, electromigration in the routing stage is difficult. So, we are applying NDR rules (double space, double-width) before routing.

• **Ques: 37. What are the types of blockages?**

Solution:

Hard blockage:- It does not allow inverters, buffers, standard cells.

Soft blockage:- It allows only inverters and buffers and blocks standard cells.

Partial blockage:- It will allow both buffers and standard cell in a percentage value.

• **Ques: 38. What is congestion?**

Solution:

When the available tracks are less than the required tracks this effect will occur. When the signals are more than the tracks then congestion will occur.

• **Ques: 39. How to fix congestion?**

Solution:

Congestion driven placement
    Adjust cell density in the congested area (high cell density causes congestion).
    Use proper blockage
    Modify the floor plan design

• **Ques: 40. What are the types of physical verification?**

Solution:

LVS (layout vs schematic)
DRC (design rule constraint check)
ERC (electric rule check)
LEC (logical equivalence check)

• **Ques: 41. How to fix setup and hold violations at a time?**

Solution:

It is not possible to fix both at a time because if we increase the delay in data path it's good for hold and bad for setup. But there is only one way to fix it.
    Buffer the data path for hold fix.
    Slow the clock frequency for setup fix (this is not a valid fix, but we don't have other options).

• **Ques: 42. How can you avoid cross-talk?**

Solution:

 (a) Increase the spacing between the aggressor and victim nets.

(b) Shielding
(c) Maintain a stable supply
(d) Increase the drive strength of the cell
(e) Layer jumping
(f) Victim net width increasing then resistance decreases.
(g) Guard ring
(h) Cell sizing (up-sizing)

- **Ques: 43. What is scan chain reordering?**

Solution:

It is the process of reconnecting the scan chains in the design to optimize for routing by reordering the scan chain connection which improves timing and congestion.

- **Ques: 44. What is the concept of rows in the floor plan?**

Solution:

The std-cells in the design are placed in rows. All rows have equal height and spacing. The width of the row can vary. The std-cell in the row gets the power and ground connection from vdd and vss rails. Sometimes, technology allows the rows to be flip. So they can share the power and ground rails in the vdd-vss-vdd patron.

- **Ques: 45. What are the advantages of NDR's?**

Solution:

(a) By applying the double-width we can avoid EM.
(b) By applying double spacing we can avoid the cross-talk.
(c) Help's to avoid congestion at the lower metal layer.
(d) Help's pin accessibility of std-cells.

- **Ques: 46. What is temperature inversion?**

Solution:

At higher CMOS technologies cell delay increases when the temperature increases. But when you are in lower technologies i.e. below 65nm cell delay has inversely proportional to temperature.

- **Ques: 47. In reg to reg path if you have setup problem where will you insert buffer?**

Solution:

We can insert buffer near to launch flop which decreases the transition time.

Hence decreasing the wire delay therefore overall delay will decrease. When arrival time will decrease setup violations will reduce (required time-arrival time).

- **Ques: 48. What is partitioning?**

Solution:

It is the process of dividing the chip into small blocks this is done mainly to separate different functional blocks and also make placement routing easier.

- **Ques: 49. Why double via insertion?**

Solution:

To reduce the yield loss due to via failures, double via's are inserted traditionally double via's where inserted in the post route and then modify the routing to fix any DRC's.

- **Ques: 50. What is metal fill insertion?**

Solution:

At the time of etching, they use some type of chemicals due to that chemical metal loss will be more for that reaction we are inserting the metal fills.

- **Ques: 51. What is metal slotting?**

Solution:

It is the Technic for avoiding problems like metal lift-off and metal erosion.

- **Ques: 52. What are the power dissipation components?**

Solution:

Dynamic power consumption: - This occurs when signals which go through the CMOS circuit change their logic state by charging-discharging of o/p node capacitor.

Static (leakage power consumption):- It is the power consumed by the subthreshold currents and by reverse-biased diodes in a CMOS transistor.

Short circuit power consumption:- It occurs during switching on both the NMOS and PMOS transistors in the circuit and they conduct simultaneously for a short amount of time.

- **Ques: 53. What is the dishing effect?**

Solution:

It is defined as the difference between the height of the oxide in the spaces and that of the metal in the trenches. It is caused by CMP. It may reduce by some dummy fill Technics effectively.

- **Ques: 54. What is CMP (chemical mechanical polishing)?**

Solution:

It is the process of a smoothing surface with a combination of chemical and mechanical forces. It is used in IC fabrication to get a high level of polarization.

- **Ques: 55. What is the use of placement blockage?**

Solution:

(a) Defines the std-cell and macro area
(b) Reserve channels for buffer insertion
(c) Prevent cells from being placed at or near macros
(d) Prevent congestion near macros

- **Ques: 56. What are the types of global routing?**

Solution:

(a) Time-driven global routing
(b) Cross-talk driven global routing
(c) Incremental global routing

- **Ques: 57. What are the violations solved in LVS?**

Solution:

(a) Shorts
(b) Opens
(c) Missing text layers
(d) Missing lib in GDS
(e) Missing soft layers

- **Ques: 58. How to fix setup and hold violations?**

Solution:

Setup:-
Reduce the number of buffers in the path
  Replace buffers with 2 inverters
  Replace HVT cells with LVT cells
  Increase the drive size/strength
  Insert repeaters.
  Adjust the cell position in the layout.

Hold:-

By adding delay in the data path
   Decrease the drive strength in the data path.

- **Ques: 59. What are the inputs of the floor plan?**

Solution:

.v
.lib and .lef
.sdc
tlu+ file
Physical partitioning information of design
Floor plan parameters like height, width, aspect ratio, utilization.
Pin/pad position.

- **Ques: 60. What are the outputs of the floor plan?**

Solution:

Die/block area
I/O pad placed
Macro placed
Power grid design
Power pre routing
Std-cell placement area

- **Ques: 61. What is the keep-out margin?**

Solution:

It is the region around the boundary of fixed macros in design in which no other macros or standard cells not allows. It allows only buffers and inverters in their area.

- **Ques: 62. How will you synthesize the clock tree?**

Solution:

(a) Single clock-normal synthesis and optimization.
(b) Multiple clocks-synthesis each clock separately.
(c) Multiple clocks with domine crossing synthesis each clock separately and balance the skew.

- **Ques: 63. What is IR drop?**

Solution:

Each metal layer has a resistance value. When the current flows through the metal the resistance consumes some current. This is the IR drop. If the resistance is more, the drop will also be more.

- **Ques: 64. How to reduce power dissipation using HVT and LVT in the design?**

Solution:

If we have positive slack use HVT cells in the path and use LVT cells in the path when we have negative slack. HVT cells have a large delay and less leakage power. LVT cells have less delay and more leakage power. To meet the timing, use LVT cells and to reduce the leakage power use HVT cells.

- **Ques: 65. What is the wire load model (WLM)?**

Solution:

It is an estimation of delay based on area and fan-out. The delay depends on the following constraints:

Resistance
Capacitance
Area of the nets

- **Ques: 66. What is signal integrity?**

Solution:

An electric signal can carry information reliably and to resist the effects (cross-talk, EM) of the high-frequency electromagnetic interface from nearby signals.

- **Ques: 67. Does cross-talk always cause violations?**

Solution:

Yes, it is because cross-talk adds or subtracts energy to the signal which causes setup or hold time violations.

- **Ques: 68. How a positive or negative edge triggered flip flop will affect the setup and hold violations?**

Solution:

Positive edge triggered flip flop will favor to setup (setup violations will reduce).

The negative edge-triggered flip flop will favor holding (hold violations will reduce).

- **Ques: 69. What are the i/p's and o/p's of power planing?**

Solution:

Inputs:-

Database with a valid floor plan.
Power rings and power straps width.
Spacing between vdd and vss straps.

Outputs:-

Design with the power structure.

- **Ques: 70. What are the i/p's and o/p's of placement?**

Solution:

Inputs:-

Netlist
Mapped and floor planed design
Logical and physical lib
Design constraints.

Outputs:-

Physical layout information
Cell placement location
Physical layout, timing, and technical information of lib

- **Ques: 71. If we increase the fan-out of the cell how it will affect delay?**

Solution:

Fan-out leads to increased capacitive load on the driving gate. Therefore, longer propagation delay.

- **Ques: 72. What is magnetic placement?**

Solution:

To improve the timing for the design or to improve the congestion for a complex floor plan we can use magnetic placement to specify fixed objects like magnets and Icc moves their connected standard cells close to them. For the best results perform the magnetic placement before standard cells are placed.

- **Ques: 73. What is the lookup table?**

Solution:

The table is drawn by using input transition and output load values. It is used to calculate the cell delay.

- **Ques: 74. What do we do for low power design?**

Solution:

Apply the low power techniques such as:
Clock gating.
Multi-voltage design.
Power gating.
Multiple vt libraries.

- **Ques: 75. What are the types of checks done in prime time?**

Solution:

(a) Timing (setup, hold, transition).
(b) Design constraints.
(c) Nets.
(d) Noise.
(e) Clock skew.

- **Ques: 76. What analysis we do during the floor plan?**

Solution:

(a) Overlapping of macros.
(b) Allowable IR drop.
(c) Global route congestion.
(d) Physical information of the design.

- **Ques: 77. What are the different types of delay models?**

Solution:

WLM (wire load model)
NLDM (Nonlinear delay model)
CCS (composite current source)

- **Ques: 78. Why we apply NDR's in placement?**

Solution:

Applying NDR's in placement because of avoiding congestion and timing problems. These problems are difficult to fix at routing. These are special rules like double spacing and double width.

- **Ques: 79. What is the mesh?**

Solution:

The horizontal and vertical power straps in the design are called mesh.

- **Ques: 80. Why I/O cells are placed in the design?**

Solution:

The i/o cells are the one which interacts in between the blocks outside of the chip to internal blocks of the chip. In-floor plan stage i/o cells are placed in between the core and die. These are responsible for providing voltage to the cell in the core.

- **Ques: 81. What are the complex cells in the floor plan?**

Solution:

These are the cells which are made of a group of std-cells based on functionality requirement. The height of this cell is greater than the std-cells and lesser than the macros.

- **Ques: 82. How to fix Electromigration (EM)?**

Solution:

(a) Downsize the driver.
(b) Increase the metal width.
(c) Add more vias.
(d) Spread cells.

- **Ques: 83. What is SOI technology?**

Solution:

It refers to the use of a layered silicon insulator. It reduces leakage current and lowers power consumption.

- **Ques: 84. What are aggressor and victim?**

Solution:

These two terms will come in the cross-talk concept.

Aggressor:- A net that creates an effect on the nearer net(victim).

Victim:- A net which receives the effect from the nearer net(aggressor).

References

1. Sherwani, N. A. (2012). *Algorithms for VLSI physical design automation*. Springer Science & Business Media.
2. https://vlsibasic.blogspot.com/2014/01/
3. Sait, S. M., & Youssef, H. (1999). *VLSI physical design automation: theory and practice* (Vol. 6). World Scientific Publishing Company.
4. Alpert, C. J., Mehta, D. P., & Sapatnekar, S. S. (2008). *Handbook of algorithms for physical design automation*. Auerbach Publications.
5. Lim, S. K. (2008). *Practical problems in VLSI physical design automation*. Springer Science & Business Media.

4

Testing of VLSI Circuits

4.1 Introduction

Testing is an integral part of the VLSI design cycle. With the advancement in IC technology, designs are becoming more and more complex, making their testing challenging. Testing occupies 60–80% time of the design process. A well-structured method for testing needs to be followed to ensure high yield and proper detection of faulty chips after manufacturing.

With the term Integrated Circuit (IC) or VLSI Testing, we refer to those procedures that take place after chip fabrication to detect possible manufacturing defects. C testing has evolved from the patterns established some years ago in the production of semiconductors. Since manual testing cannot meet the complex needs indigenous to IC manufacture, highly sophisticated instruments, and test systems have developed that are automatically programmed by computer, tape, or printed-circuit cards.

4.1.1 Rule of Ten

The earlier a defect is detected, the lesser the cost of the final product. The rule of ten says that the cost of detecting a defective device increases by an order of magnitude as we move from a manufacturing stage to the next (device→board→system)

4.2 Testing of a Circuit

The testing of a circuit is the concept of applying a set of test stimuli to the circuit under test and observe the results. It includes:

- Inputs of the circuit under test (CUT), and
- Analyzing output responses
- If incorrect (fail), CUT assumed to be faulty and if correct (pass), CUT assumed to be fault-free. The VLSI design cycle is shown in Figure 4.1 and design for testability is shown in Figure 4.2.

93

Figure 4.1   VLSI testing life cycle.

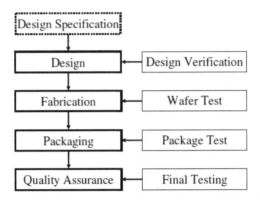

Figure 4.2   Different phase of testing during VLSI cycle.

4.2.1 Testing During VLSI Development

Semiconductor test equipment (IC tester), or automated test equipment (ATE), is a system for giving electrical signals to a semiconductor device to compare output signals against expected values for testing if the device works as specified in its design specifications.

Testers are roughly categorized into logic testers, memory testers, and analog testers. Normally, IC testing is conducted at two levels: the wafer test (also called die sort or probe test) that tests wafers, and the package test (also called final test) after packaging. Wafer testing uses a prober and a probe card, while package testing uses a handler and a test socket, together with a tester.

4.2.1.1 Yield

The manufacturing defects lead to faulty chips. The yield can be calculated as:

$$Yield = \frac{Number\ of\ accptable\ components}{Total\ number\ of\ fabricated\ components}$$

There are two types of yield losses:

1. Catastrophic: due to random defects
2. Parametric: due to process variations

4.2.2 Design For Test (DFT)

Design for testing or design for testability (DFT) consists of IC design techniques that add testability features to a hardware product design. The added features make it easier to develop and apply manufacturing tests to the designed hardware. The purpose of manufacturing tests is to validate that the product hardware contains no manufacturing defects that could adversely affect the product's correct functioning.

4.2.2.1 Automatic Test Pattern Generation (ATPG)

ATPG (acronym for both Automatic Test Pattern Generation and Automatic Test Pattern Generator) is an electronic design automation method/technology used to find an input (or test) sequence that, when applied to a digital circuit, enables automatic test equipment to distinguish between the correct circuit behavior and the faulty circuit behavior caused by defects. The generated patterns are used to test semiconductor devices after manufacture or to assist with determining the cause of failure (failure analysis).

4.2.2.2 Defect and Error

Testing is the process of identifying defects, where a defect is any variance between actual and expected output. A mistake in HDL coding is called error. The error found by the tester is called a defect. The product or design does not meet the requirements then it is a failure. So, a defect is an error caused in a device during the manufacturing process.

4.2.3 Fault Model

A fault model is a mathematical description of how a defect alters design behaviour. The logic values observed at the device's primary outputs, while applying a test pattern to some device under test (DUT), are called the output of that test pattern. The output of a test pattern, when testing a fault-free device that works exactly as designed, is called the expected output of that test pattern.

Basic fault models in digital circuits include:

- The stuck-at fault model.
  A signal, or gate output, is stuck at a 0 or 1 value, independent of the inputs to the circuit.

- The bridging fault model.
  Two signals are connected when they should not be. Depending on the logic circuitry employed, this may result in a wired-OR or wired-AND logic function.
- The transistor faults.
  This model is used to describe faults for CMOS logic gates. At the transistor level, a transistor may be stuck-short or stuck-open. In stuck-short, a transistor behaves as it is always conducting (or stuck-on), and stuck-open is when a transistor never conducts current (or stuck-off). Stuck-short will produce a short between VDD and VSS.
- The open fault models.
  A wire is assumed broken, and one or more inputs are disconnected from the output that should drive them. As with bridging faults, the resulting behaviour depends on the circuit implementation.
- The delay fault model
  The signal eventually assumes the correct value, but more slowly (or rarely, more quickly) than normal.

4.2.3.1 Detection of fault

A fault is said to be detected by a test pattern if the output of that test pattern when testing a device that has only that one fault, is different than the expected output.

4.2.3.2 Phases of fault

The ATPG process for a targeted fault consists of two phases: fault activation and fault propagation. Fault activation establishes a signal value at the fault model site that is opposite of the value produced by the fault model. Fault propagation moves the resulting signal value, or fault effect, forward by sensitizing a path from the fault site to a primary output.

Technical Questions with Solutions

- **Ques: 1. What do you understand by VLSI Testing?**

Solution:

Testing is a process that includes test pattern generation, test pattern application, and output evaluation. A well-structured method for testing needs to be followed to ensure high yield and proper detection of faulty chips after manufacturing.

If N is the number of transistors and P is the probability that the transistor is faulty, then the probability that chip is faulty is: $1 - (1 - P)^N$

- **Ques: 2. Mention the levels at which testing of a chip can be done?**

Solution:

(a) At the wafer level
(b) At the packaged chip level
(c) At the board level
(d) At the system level
(e) In the field

- **Ques: 3. What are the different categories of testing?**

Solution:

(a) Functionality tests
(b) Manufacturing tests

- **Ques: 4. Write notes on functionality tests?**

Solution:

Functionality tests verify that the chip performs its intended function. These tests assert that all the gates in the chip, acting in concert, achieve the desired function. These tests are usually used early in the design cycle to verify the functionality of the circuit.

- **Ques: 5. Write notes on manufacturing tests?**

Solution:

Manufacturing tests verify that every gate and register in the chip functions correctly. These tests are used after the chip is manufactured to verify that the silicon is intact.

- **Ques: 6. Mention the defects that occur in a chip?**

Solution:

(a) layer-to-layer shorts
(b) discontinuous wires
(c) thin-oxide shorts to substrate or well
(d) Oxide breakdown
(e) Electromigration, which is primarily caused by the transport of metal atoms when a current flow through the wire.

Because of a low melting point, aluminum has large self-diffusion properties, which increase its electromigration liability.

• **Ques: 7. Give some circuit maladies to overcome the defects?**

Solution:

(a) nodes shorted to power or ground
(b) nodes shorted to each other
(c) inputs floating/outputs disconnected

• **Ques: 8. What are the tests for I/O integrity?**

Solution:

(a) I/O level test
(b) Speed test
(c) IDD test

• **Ques: 9. What is meant by fault models?**

Solution:

The fault model is a model for how faults occur and their impact on circuits.

• **Ques: 10. Give some examples of fault models?**

Solution:

(a) Stuck-At Faults
(b) Short-Circuit and Open-Circuit Faults

• **Ques: 11. What is stuck – at fault?**

Solution:

The given line has a constant value (0/1) independent of other signal values in the circuit.

With this model, a faulty gate input is modeled as a "stuck at zero" or "stuck at one". These faults most frequently occur due to thin oxide shorts or metal-to-metal shorts.

Properties:

• Only one line is faulty
• The faulty line is permanently set to 0 or 1
• The fault can be at an input or output of a gate
• The simple logical model is independent of technology details.
• It reduces the complexity of fault-detection algorithms.

• **Ques: 12. What is meant by observability?**

Solution:

The observability of a particular internal circuit node is the degree to which one can observe that node at the outputs of an integrated circuit.

- **Ques: 13. What is meant by controllability?**

Solution:

The controllability of an internal circuit node within a chip is a measure of the ease of setting the node to a 1 or 0 state.

- **Ques: 14. What is known as percentage-fault coverage?**

Solution:

The total number of nodes that, when set to 1 or 0, does result in the detection of the fault, divided by the total number of nodes in the circuit, is called the percentage-fault coverage.

- **Ques: 15. What is fault grading?**

Solution:

Fault grading consists of two steps. First, the node to be faulted is selected. A simulation is run with no faults inserted, and the results of this simulation are saved. Each node or line to be faulted is set to 0 and then 1 and the test vector set is applied. If and when a discrepancy is detected between the faulted circuit response and the good circuit response, the fault is said to be detected and the simulation is stopped.

- **Ques: 16. Mention the ideas to increase the speed of fault simulation?**

Solution:

(a) Parallel simulation
(b) Concurrent simulation

- **Ques: 17. What is fault sampling?**

Solution:

An approach to fault analysis is known as fault sampling. This is used in circuits where it is impossible to fault every node in the circuit. Nodes are randomly selected and faulted. The resulting fault detection rate may be statistically inferred from the number of faults that are detected in the fault set and the size of the set. The randomly selected faults are unbiased. It will determine whether the fault coverage exceeds the desired level.

- **Ques: 18. What are the approaches in design for testability?**

Solution:

(a) Ad hoc testing
(b) Scan-based approaches
(c) Self-test and built-in testing

- **Ques: 19. Mention the common techniques involved in ad hoc testing?**

Solution:

(a) Partitioning large sequential circuits
(b) Adding test points
(c) Adding multiplexers
(d) Providing for easy state reset

- **Ques: 20. What are the scan-based test techniques?**

Solution:

(a) Level sensitive scan design
(b) Serial scan
(c) Partial serial scan
(d) Parallel scan

- **Ques: 21. What are the two tenets in LSSD?**

Solution:

(a) The circuit is level-sensitive.
(b) Each register may be converted to a serial shift register.

- **Ques: 22. What are the self-test techniques?**

Solution:

(a) Signature analysis and BILBO
(b) Memory self-test
(c) Iterative logic array testing

- **Ques: 23. What is known as BILBO?**

Solution:

Signature analysis can be merged with the scan technique to create a structure known as BILBO- for Built-In Logic Block Observation.

- **Ques: 24. What is known as IDDQ testing?**

Solution:

A popular method of testing for bridging faults is called IDDQ or current-supply monitoring. This relies on the fact that when a complementary CMOS logic gate is not switching, it draws no DC. When a bridging fault occurs, for some combination of input conditions a measurable DC IDD will flow.

- **Ques: 25. What are the applications of chip-level test techniques?**

Solution:

(a) Regular logic arrays
(b) Memories
(c) Random logic

- **Ques: 26. What is a boundary scan?**

Solution:

The increasing complexity of boards and the movement to technologies like multichip modules and surface-mount technologies resulted in system designers agreeing on a unified scan-based methodology for testing chips at the board. This is called a boundary-scan.

- **Ques: 27. What is the test access port?**

Solution:

The Test Access Port (TAP) is a definition of the interface that needs to be included in an IC to make it capable of being included in boundary-scan architecture. The port has four or five single bit connections, as follows:

(a) TCK(The Test Clock Input)
(b) TMS(The Test Mode Select)
(c) TDI(The Test Data Input)
(d) TDO(The Test Data Output)
    It also has an optional signal
(e) TRST*(The Test Reset Signal)

- **Ques: 26. What are the contents of the test architecture?**

Solution:

The test architecture consists of:

(a) The TAP interface pins
(b) A set of test-data registers
(c) An instruction registers
(d) A TAP controller

- **Ques: 27. What is the TAP controller?**

Solution:

The TAP controller is a 16-state FSM that proceeds from state to state based on the TCK and TMS signals. It provides signals that control the test data registers, and the instruction register. These include serial-shift clocks and update clocks.

- **Ques: 28. What is known as the test data register?**

Solution:

The test-data registers are used to set the inputs of modules to be tested and to collect the results of running tests.

- **Ques: 29. What is known as a boundary scan register?**

Solution:

The boundary scan register is a special case of a data register. It allows circuit-board interconnections to be tested, external components tested, and the state of chip digital I/Os to be sampled.

- **Ques: 30. Circuits should be tested at which level?**

Solution:

The chip design mistakes can be very costly both in terms of time and money. The circuit should be tested at the chip-level itself. Design for testability is essential for good design.

- **Ques: 31. Draw the VLSI design cycle?**

Solution: In Figure 4.3, the VLSI design cycle is shown.

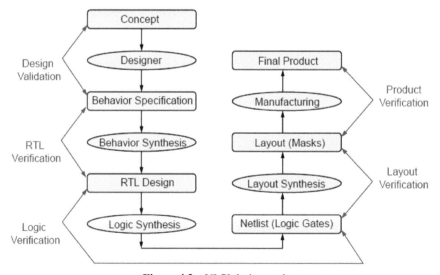

Figure 4.3   VLSI design cycle.

• **Ques: 32. What is the difference between testing and diagnosis?**

Solution:

The role of testing is to detect whether something went wrong and the role of diagnosis is to determine exactly what went wrong.

• **Ques: 33. Draw the DFT (Design For Testability) cycle?**

Solution:

DFT is Design the chip to increase observability and controllability. DFT is a technique, which facilitates a design to become testable after fabrication, as shown in Figure 4.4. "Extra" logic that we put along with the design logic during the implementation process helps post-production testing.

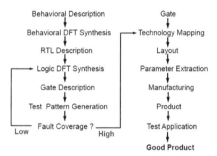

Figure 4.4   Design for Testability.

• **Ques: 34. What do you understand by verification and test?**

Solution:

Verification:

1. Verifies correctness of design
2. Performed by simulation, hardware emulation, or formal methods
3. Perform once before manufacturing
4. Responsible for quality of design

Test:

1. Verifies correctness of manufactured hardware
2. It is a two-part process

    (a) Test generation: software process executed once during design
    (b) Test application: electrical tests applied to hardware

3. Test application performed on every manufactured device
4. Responsible for quality of the device

- **Ques: 35. Show the Verification and test procedure by the Re-convergent path model?**

Solution: In following Figure 4.5 shows the re-convergent path model for verification.

Figure 4.5 Re-convergent path model.

- **Ques: 36. Differentiate between defect, fault, and error?**

Solution:

Defect:

A defect is an unintended difference between the implemented hardware and its intended design.

The Defects occur either during manufacture or during the use of devices

Fault:

It is a representation of a defect at the abstracted function level

Error:

A wrong output signal produced by a defective system.

An error is caused by a Fault or a design error

- **Ques: 37. Explain defect, fault, and error with an example of AND gate.**

Solution:

Consider one two-input AND gate, as in Figure 4.6.

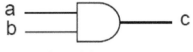

Figure 4.6 AND gate.

- Defect: a short to ground, shown in Figure 4.7.

Figure 4.7 Defect in a logic gate.

- Fault: signal b stuck at logic 0
- Error: a = 1, b = 1, c = 0 (correct output c = 1)

The error is not permanent. As long as at least one input is 0, there is no error in the output.

- **Ques: 38. How a test problem is treated?**

Solution: The design flow for test and fault is shown in Figure 4.8.

Figure 4.8 Fault detection.

- **Ques: 39. What is the 'Rule of ten' in chip testing?**

Solution:

Chips must be tested before they are assembled onto PCBs, which, in turn, must be tested before they are assembled into systems.

If a chip fault is not detected by chip testing, then finding the fault costs 10 times as much at the PCB level as at the chip level.

Similarly, if a board fault is not found by PCB testing, then finding the fault costs 10 times as much at the system level as at the board level.

- **Ques: 40. How to calculate the cost of the chip?**

Solution:

Fraction (or percentage) of good chips produced in a manufacturing process is called the yield. Yield is denoted by symbol Y.

The Cost of a chip is:

$$\frac{\text{Cost of fabricating and testing a wafer}}{\text{Yield} \times \text{Number of chip sites on the wafer}}$$

- **Ques: 41. What is fault coverage?**

Solution:

The fault coverage is the measure of the ability of a test (a collection of test patterns) to detect a given fault that may occur on the device under test.

$$\text{Fault Coverage} = \frac{\text{Detected Faults}}{\text{Possible Faults}}$$

- **Ques: 42. What do you understand by the detection limit?**

Solution:

The ratio of faulty chips among the chips that pass tests

- DL is measured as defects per million (DPM)
- DL is a measure of the effectiveness of tests
- DL is a quantitative measure of the manufactured product quality. For commercial VLSI chips, a DL greater than 500 DPM is considered unacceptable.

$$DL = 1 - Y^{(1-FC)}$$

- The detection level should be in between 0 and Y is:

$$0 < DL \leq 1 - Y$$

- **Ques: 45. What is the Quality level (QL) and what is the relation between QL and DL?**

Solution:

The fraction of good parts among the parts that pass all the tests are shipped.

$$QL = 1 - DL$$

- **Ques: 46. What are the different categories of defects?**

Solution:

(a) Random defects, which are independent of designs and processes
(b) Systematic defects, which depend on designs and processes used for manufacturing

For example, random defects might be caused by random particles scattered on a wafer during manufacturing. Systematic defects might be caused by process variations, signal integrity, and design integrity issues.

- **Ques: 47. What do you mean by logical fault?**

Solution:

Logical faults represent the physical defects in the behaviors of the systems.

- **Ques: 48. Which are the common faults modeled in Transistors?**

Solution:

MOS transistor is considered an ideal switch and two types of faults are modeled.

1. Stuck-open – a single transistor is permanently stuck in the open state

   - Turn the circuit into a sequential one
   - Need a sequence of at least 2 tests to detect a single fault.
   - Unique to CMOS circuits

2. Stuck-on – a single transistor is permanently shorted irrespective of its gate voltage

Detection of a stuck-open fault requires two vectors.

Detection of a stuck-short fault requires the measurement of quiescent current (IDDQ).

- **Ques: 49. What are fault detection, fault location, and fault diagnosis?**

Solution:

Fault detection tells whether a circuit is fault-free or not.
   The fault location provides the location of the detected fault.
   Fault diagnosis provides the location and the type of the detected fault.

- **Ques: 50. What is the role of a Built-In Self Test (BIST)?**

Solution:

Built-in self-test lets blocks test themselves

- Generate pseudo-random inputs to combinational logic.
- Combine outputs into a syndrome.
- With high probability, the block is fault-free if it produces the expected syndrome.

- **Ques: 51. Give the flow graph of the ATPG (Automatic Test Pattern Generator)?**

Solution: The Figure 4.9 shows the flow graph of ATPG.

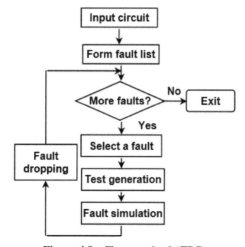

Figure 4.9   Flow graph of ATPG.

References

1. http://www.ee.ncu.edu.tw/~{ }jfli/vlsi21/lecture/ch06.pdf
2. Girard, P. (2002). Survey of low-power testing of VLSI circuits. *IEEE Design & test of computers*, *19*(3), 82–92.
3. Krstic, A., & Cheng, K. T. T. (2012). *Delay fault testing for VLSI circuits* (Vol. 14). Springer Science & Business Media.
4. Nicolici, N., & Al-Hashimi, B. (2003). *Power-constrained testing of VLSI circuits* (p. 178). Boston, MA: Kluwer Academic Publishers.
5. https://en.wikipedia.org/wiki/Design_for_testing

5

Miscellaneous

Electronics and Electrical engineering is an emerging and most popular engineering discipline. Electronics engineering and electrical engineering are closely related to this field. Even imagining a life bereft of electronic gadgets seems impossible in today's world. There is no field left across the globe where one cannot find the usage of electronics and communication engineering. Perhaps that is why electronics have become the vertebrae of digital technology.

5.1 Branches of Electronics

Electronics is further having emerging branches which target various industry and interest.

5.1.1 Electronics and Communication Engineering (ECE)

This branch deals with analog and digital transmission & reception of data, networking, voice and video, solid-state devices, microprocessors, digital and analog communication, satellite communication, antennae, and wave progression. It also deals with the manufacturing of electronic devices, circuits, and communications equipment like transmitter, receiver, integrated circuits, microwaves, and fiber among others.

An electronics engineer is qualified for jobs that include building electronic components for integration into larger systems, Integrated Circuit (IC) design work, board layout, programming microcontrollers and microprocessors, and field testing.

Electronics and communication engineers deal with all of the applications which make our life easier and enjoyable such as television, radio, computers, mobiles, etc. are designed and developed by Electronics and Communication Engineers.

5.1.2 Electronics and Telecommunication Engineering (ETE)

This branch deals with various principles and practical aspects related to designing various telecommunication equipment. Electronics and Telecommunication Engineers develop prototypes of integrated circuit components. Apart from this, it also combines various aspects of electrical, structural and civil engineering as well.

Telecommunications engineers handle different types of technology that help us to communicate. They research, design and develop satellite and cable systems, mobile phones, radio waves, the internet, and e-mail.

Electronics and telecommunication engineers deal with consumer electronics, aviation and avionics, manufacturing, electricity generation and distribution, communications, transportation, telecommunications industry as electronics engineers.

5.1.3 Microelectronics and VLSI Design

Microelectronics is a field of specialization and VLSI is a process under this specialization. VLSI design deals with "design for VLSI", it could be analog, digital, mixed-signal, etc, which is again a process under Microelectronics. Microelectronics tends to focus on a circuit that is developed for a specific purpose, while very large systems integration (VLSI) tends to focus on integrated many circuits to perform a more difficult or more general purpose.

An example of a microelectronic would be an accelerometer that uses electronic tunneling to output a voltage based on acceleration. An example of a VLSI would be the microcontroller that is used to interpret and utilize the accelerometer signal.

Technical Questions with Solutions

• **Ques: 1. What are the advantages of VLSI Design?**

Solution:

The number of transistors in an IC has dramatically increased.

VLSI lets all these into one chip resulting in the following advantages:

1. Reduces the size of circuits.
2. Reduces the cost of the devices.
3. Increases operating speed of circuits.

• **Ques: 2. What is the future of VLSI technology?**

Solution:

The future of VLSI circuits depends on the trend of channel length reduction. Available fabrication technologies deny more degradation in channel length, so nanoelectronics devices such as Quantum-dot Cellular Automata (QCA), Single Electron Transistor (SET), Carbon nanotube field-effect transistors (CNTFET) and benzene rings are candidates for replacement of conventional CMOS technology.

• **Ques: 3. What is the transistor count in the latest VLSI technology?**

Solution:

The transistor count is the number of transistors on an integrated circuit (IC). The rate at which transistor counts have increased generally follows Moore's law, which observed that the transistor count doubles approximately every two years.

In the latest 7nm process technology, the APPLE has released Apple A12X Bionic Processor (Octa-core AMD 64 mobile SOC) which has 10,000,000,000 transistors in $122m^2$ area.

• **Ques: 4. What is the leakage current in CMOS?**

Solution:

The leakage current is an unwanted conductive path under normal operating conditions. When Vgs<Vth, the current leaks between drain and source of MOS. The acceptance value of leakage current is 210ţA.

There are four main leakage currents in a CMOS transistor, as shown in Figure 5.1.

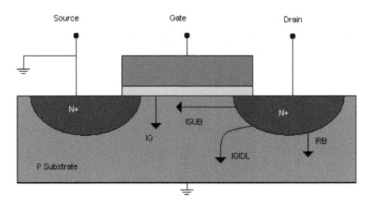

Figure 5.1 CMOS transistor.

A. Subthreshold (weak inversion) leakage (ISUB)

The subthreshold leakage current is a drain to source current of a transistor when it operates in the weak inversion region. It is due to the diffusion current of minority carriers in the channel of a MOS structure. When the gate voltage is below Vth, the turn-on or threshold voltage, an NMOS device is turned off.

B. Reverse-biased Source/Drain junction leakage (IRB)

Recalling the structure of an NMOS transistor, a reverse-biased p-n junction exists that is formed by either the source or drain to the substrate when it is off. Although a potential barrier exists, there is a leakage current. It has two components: electron-hole generation in the depletion region and minority carrier diffusion/drift at the edge of the same depletion region.

C. Gate direct-tunneling leakage (IG)

The current due to gate direct-tunneling leakage is current tunneling into the gate of the transistor. The mechanisms for this tunneling effect are electron conduction band tunneling (ECB), electron valence band tunneling (EVB), and hole valence band tunneling (HVB). Currently, ECB is the dominant phenomenon. There is electron tunneling because of the high electric fields that exist across the oxide layer

D. Gate-induced drain leakage (IGIDL)

The current due to gate-induced drain leakage (GIDL) is a direct consequence of the high field effect in the drain of a MOS transistor. For example, an NMOS with its gate grounded and drain at the supply voltage potential experiences energy band bending in the drain. This allows electron-hole pairs to be generated by avalanche effects and band-to-band carrier tunneling. Holes are quickly driven to the body, creating a deep depletion situation. Similarly, electrons are collected in the drain. Together, these effects create the IGIDL current.

- **Ques: 5. What do you mean by parasitic capacitance in MOSFET?**

Solution:

These are unwanted capacitances, but still are part of the transistor. Together with the resistances in the circuit, they put an upper limit to the speed of the transistor.

When two electrical conductors at different voltage are close together, the electric field between them causes electric charge to be stored on them. This is known as the parasitic capacitance effect.

Types of parasitic capacitances

1. Junctions depletion capacitances
2. Overlap and gate-channel capacitances
3. Channel-bulk depletion capacitance

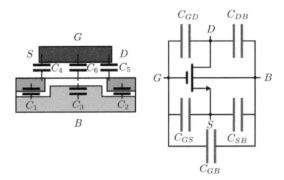

Figure 5.2 Parasitic capacitances.

The Figure 5.2 shows various parasitic capacitances. C1 and C2 are capacitances created by the depletion regions between the source/drain and bulk. C3 is the depletion capacitance between the channel and bulk. C4 and C5 are capacitances caused by the overlap between the gate and the source/drain diffusions. Finally, C6 is the oxide capacitance between the gate and the channel and is split between drain and source depending on the region of operation of the transistor.

- **Ques: 6. What do you mean by FINFET?**

Solution:

FinFET, also known as Fin Field Effect Transistor, as shown in Figure 5.3, is a type of non-planar or "3D" transistor used in the design of modern processors. As in earlier, planar designs, it is built on an SOI (silicon on insulator) substrate. However, FinFET designs also use a conducting channel that rises above the level of the insulator, creating a thin silicon structure, shaped like a fin, which is called a gate electrode. This fin-shaped electrode allows multiple gates to operate on a single transistor. Intel began releasing FinFET CPU technology in 2012 with its 22-nm Ivy Bridge processors.

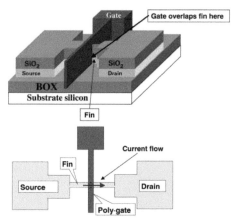

Figure 5.3   FINFET.

Dr. Chenming Hu has been called the Father of 3D Transistor for developing the FinFET in 1999.

- **Ques: 7. What are the benefit and drawbacks of FINFET?**

Solution:

- To exploit different benefits of FinFET, it is fabricated into two types:
    (1) Dual-gate FinFET, shown in Figure 5.4, which trims the excess silicon by fabricating the channel using an ultra-thin layer of silicon that sits on top of an insulator, therefore the electric field from the gate to the fin on the top is drastically reduced.

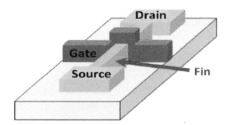

Figure 5.4   Dual-gate FINFET.

    (2) tri-gate FinFET, shown in Figure 5.5, in which the FET gate wraps around three sides of the transistor's elevated channel, or "fin". Since fins are made vertical, high packing density can be achieved,

by packing transistors closer together. Further, to get even more performance and energy-efficiency gains, designers also can continue
growing the height of the fins.

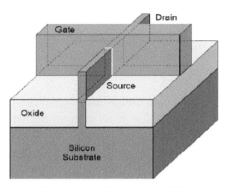

Figure 5.5 Tri-gate FINFET.

- It exhibits little or no body effect because FinFET channels are fully
  depleted.
- Given the excellent control of the conducting channel by the gate, very
  little current is allowed to leak through the body when the device is in
  the off state. The FinFET can also be run at a lower operating voltage for
  given leakage current, halving its dynamic power consumption (which
  is proportional to CV2f) for a 0.7 scaling in VDD.
- At 1V, the FinFET is 18% faster than the equivalent planar device, but
  at 0.7V, the advantage is 37%. This enables the device to be operated at
  lower threshold voltages for the same leakage.
- The difference between the gate and threshold voltage at very low
  operating voltages is much greater, thus exaggerating the performance
  advantage of very low-voltage FinFETs.
- Increased voltage headroom for circuits such as cascades
- Lower gate resistance helps keep flicker noise under control as well as
  improved matching
- Higher current drive and higher gain.

However, the disadvantages are as follows:

- The designer cannot control the channel as easily and the higher
  source/drain resistance cuts transconductance.
- Designers have little choice overvoltages for I/O and have to develop
  more complex methods to achieve ESD immunity.

- The Si surface of fins appears different than in bulk, therefore excessive Si loss was observed after the usual pre-gate-oxide clean. Thus, wet cleans are optimized with dilute concentration and lower temperatures. Similarly, the oxidation of fins is also faster at the corner and tip of fins.
- Besides, the dry etching on fins is more stringent due to the 3D structures and a bias plasma pulsing scheme may be viable for minimizing Si loss.

- **Ques: 8. What is skin effect?**

Solution:

Every conductor line has resistance and inductance! When you pass AC in a conductor (imagine a rod), then due to its alternating nature and with the conductor inductance, it finds some opposition to the flow of current in the regions closer to the center of the conductor. Since this inductive reactance is stronger at the center, the current tends to avoid that area and prefers flowing on the periphery or the skin of the conductor. Hence, skin effect. The more the frequency, the stronger is the reactance. The concept of skin effect is shown in Figure 5.6.

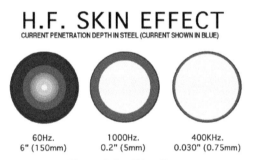

Figure 5.6   Skin effect.

- **Ques: 9. Why skin effect occurs in AC, but not in DC?**

Solution:

The skin effect is caused by the back emf produced by the self-induced magnetic flux in a conductor. For a DC, the rate of change of flux is zero, so there is no back emf due to changes in magnetic flux. Therefore, the current is uniformly distributed throughout the cross-section of the conductor.

In the case of DC, there is no change in frequency and hence the inductive reactance is zero. So, the current does not find any opposition other than the resistance and it flows completely in the conductor, not just outside.

- **Ques: 10. What do you mean by MuGFET?**

Solution:

A multigate device or multiple-gate field-effect transistor (MuGFET) refers to a MOSFET (metal-oxide-semiconductor field-effect transistor) that incorporates more than one gate into a single device. The multiple gates may be controlled by a single gate electrode, wherein the multiple-gate surfaces act electrically as a single gate, or by independent gate electrodes. A multigate device employing independent gate electrodes is sometimes called a multiple-independent-gate field-effect transistor (MIGFET).

- **Ques: 11. What is a tri-gate transistor or 3D transistor?**

Solution:

Tri-gate or 3D transistor (not to be confused with 3D microchips) fabrication is used by Intel Corporation for the nonplanar transistor architecture used in Ivy Bridge, Haswell, and Skylake processors. These transistors employ a single gate stacked on top of two vertical gates (a single gate wrapped over 3 sides of the channel), allowing essentially three times the surface area for electrons to travel. Intel reports that their tri-gate transistors reduce leakage and consume far less power than current transistors. This allows up to 37% higher speed or power consumption at under 50% of the previous type of transistors used by Intel.

- **Ques: 12. What is electronics? Define it.**

Solution:

Electronics is the branch of science that deals with the study of flow and control of electrons (electricity) and the study of their behavior and effects in vacuum, gas, and semiconductors, and with devices using such electrons. This control of electrons is accomplished by devices that resist, carry, select, steer, switch, store, manipulate, and exploit the electron.

- **Ques: 13. What is communication?**

Solution:

Communication means transferring a signal from the transmitter which passes through a medium then the output is obtained at the receiver (or) communication says as transferring a message from one place to another place called communication.

• **Ques: 14. Explain types of communication?**

Solution:

There are two types of communication: analog and digital communication.

As a technology, analog is the process of taking an audio or video signal (the human voice) and translating it into electronic pulses. On the other hand, digital is breaking the signal into a binary format where the audio or video data is represented by a series of "1"s and "0"s.

Digital signals are immune to noise, quality of transmission and reception is good, components used in digital communication can be produced with high precision and power consumption is also very less when compared with analog signals.

• **Ques: 15. What are active and passive components? Give some examples?**

Solution:

Passive: Capable of operating without an external power source. Typical passive components are resistors, capacitors, inductors, and diodes (although the latter are a special case).

Active: Requiring a source of power to operate. It includes transistors (all types), integrated circuits (all types), TRIACs, SCRs, LEDs, etc.

• **Ques: 16. What are DC and AC?**

Solution:

DC: Direct Current. The electrons flow in one direction only. Current flow is from negative to positive, although it is often more convenient to think of it as from positive to negative. This is sometimes referred to as "conventional" current as opposed to electron flow.

C: Alternating Current. The electrons cyclically flow in both directions— first one way, then the other. The rate of change of direction determines the frequency, measured in Hertz (cycles per second).

• **Ques: 17. Define the term 'frequency'?**

Solution:

Frequency: Unit is Hertz, Symbol is Hz, the old symbol was cps (cycles per second). A complete cycle is completed when the AC signal has gone from zero volts to one extreme, back through zero volts to the opposite extreme, and returned to zero. The accepted audio range is from 20Hz to 20,000Hz. The number of times the signal completes a complete cycle in one second is the frequency.

- **Ques: 18. What do you mean by 'voltage'?**

Solution:

Voltage: Unit is Volts, Symbol is V or U, the old symbol was E . Voltage is the "pressure" of electricity, or "electromotive force" (hence the old term E). A 9V battery has a voltage of 9V DC and may be positive or negative depending on the terminal that is used as the reference. The mains have a voltage of 220, 240, or 110V depending where you live - this is AC, and alternates between positive and negative values. Voltage is also commonly measured in millivolts (mV), and 1000 mV is 1V. Microvolts (UV) and nanovolts (nV) are also used.

- **Ques: 19. What do you mean by 'current'?**

Solution:

Current: Unit is Amperes (Amps), Symbol is I. Current is the flow of electricity (electrons). No current flows between the terminals of a battery or other voltage supply unless a load is connected. The magnitude of the current is determined by the available voltage and the resistance (or impedance) of the load and the power source. Current can be AC or DC, positive or negative, depending upon the reference. For electronics, current may also be measured in mA (milliamps) – 1000 mA is 1A. Nanoamps (nA) is also used in some cases.

- **Ques: 20. What do you mean by 'resistance'?**

Solution:

Resistance: Unit is Ohms, Symbol is R or Ω. Resistance is a measure of how easily (or with what difficulty) electrons will flow through the device. Copper wire has a very low resistance, so a small voltage will allow a large current to flow. Likewise, the plastic insulation has a very high resistance and prevents current from flowing from one wire to those adjacent. Resistors have a defined resistance, so the current can be calculated for any voltage. Resistance in passive devices is always positive (i.e. > 0)

- **Ques: 21. Give the SI units and symbols of various electronic components?**

Solution:

Table 5.1 Various SI symbols of electronic components

Name	Symbol	Quantity	In other SI units	In SI base units
radian	rad	plane angle	1	$(m \cdot m^{-1})$
steradian	sr	solid angle	1	$(m^2 \cdot m^{-2})$

(Continued)

Table 5.1   Continued

Name	Symbol	Quantity	In other SI units	In SI base units
hertz	Hz	frequency		s^{-1}
newton	N	force, weight		$kg \cdot m \cdot s^{-2}$
pascal	Pa	pressure, stress	N/m^2	$kg \cdot m^{-1} \cdot s^{-2}$
joule	J	energy, work, heat	$N \cdot m = Pa \cdot m^3$	$kg \cdot m^2 \cdot s^{-2}$
watt	W	power, radiant flux	J/s	$kg \cdot m^2 \cdot s^{-3}$
coulomb	C	electric charge or quantity of electricity		$s \cdot A$
volt	V	voltage (electrical potential), emf	W/A	$kg \cdot m^2 \cdot s^{-3} \cdot A^{-1}$
farad	F	capacitance	C/V	$kg^{-1} \cdot m^{-2} \cdot s^4 \cdot A^2$
ohm	Ω	resistance, impedance, reactance	V/A	$kg \cdot m^2 \cdot s^{-3} \cdot A^{-2}$
Siemens	S	electrical conductance	Ω^{-1}	$kg^{-1} \cdot m^{-2} \cdot s^3 \cdot A^2$
weber	Wb	magnetic flux	$V \cdot s$	$kg \cdot m^2 \cdot s^{-2} \cdot A^{-1}$
tesla	T	magnetic flux density	Wb/m^2	$kg \cdot s^{-2} \cdot A^{-1}$
henry	H	inductance	Wb/A	$kg \cdot m^2 \cdot s^{-2} \cdot A^{-2}$
degree Celsius	°C	temperature relative to 273.15 K		K
lumen	lm	luminous flux	$cd \cdot sr$	cd
lux	lx	illuminance	lm/m^2	$m^{-2} \cdot cd$

- **Ques: 22. Define 1 ohm?**

Solution:

One ohm is the unit of resistance. It is equal to 1 volt upon 1 ampere. It means that an object having a resistance of 1ohm allows 1 ampere of current to flow through 2 points having a potential difference of 1 volt.

• **Ques: 23. Define 1 ampere?**

Solution:

1 Ampere: One Ampere is defined as the current that flows with an electric charge of one Coulomb per second. or, the ampere is that current which, when passing through a resistance of 1 ohm, produces a potential difference of 1 V across its terminals.

• **Ques: 24. Define 1 watt?**

Solution:

If a circuit is having pure resistance and in which we apply a 1-volt voltage through which 1-ampere current is produced by load then power consumption is 1 watt.

Power is defined as the rate of doing work.

$$Power = \frac{work}{time}$$

1 watt is the power of an appliance that consumes energy at the rate of 1 joule per second.

• **Ques: 25. Define 1 Joule?**

Solution:

Joule is an SI unit of work. 1 Joule is the amount of work done when a force of 1 Newton displaces a body through a distance of 1m in the direction of the force applied.

• **Ques: 26. Define 1 volt?**

Solution:

One Volt is defined as energy consumption of one joule per electric charge of one coulomb. $1V = 1J/C$. One volt is equal to the current of 1amp times the resistance of 1 ohm: $1V = 1A \cdot 1\Omega$

One volt is the difference in electrical potential between two points of conducting wire when an electric current of 1A dissipates 1W power between those points.

• **Ques: 27. Define 1 farad?**

Solution:

One farad is defined as the capacitance across which, when charged with one coulomb, there is a potential difference of one volt.

$$C = \frac{Q}{V}$$

The 1-farad capacitor can store one Columb of charge at 1 volt.

• **Ques: 28. Define 1 columb?**

Solution:

It is the basic unit of electric charge, equal to the quantity of charge transferred in one second by a steady current of one ampere, and equivalent to 6.2415×1018 elementary charges, where one elementary charge is the charge of a proton or the negative of the charge of an electron.

Amount of charge transported by a constant current of 1 Amp in 1 second is 1 columb

$$Q = I \times T$$

• **Ques: 29. What is an electronic circuit?**

Solution:

A circuit is a structure that directs and controls electric currents, presumably to perform some useful function. The very name "circuit" implies that the structure is closed, something like a loop.

• **Ques: 30. Define 'pressure' and 'force'?**

Solution:

The pressure is an expression of force exerted on a surface per unit area. The standard unit of pressure is the pascal.

$$F = M \times A \quad \text{and} \quad P = \frac{F}{A}$$

A force is any interaction that, when unopposed, will change the motion of an object.

• **Ques: 31. Define 1 pascal and 1 newton?**

Solution:

Pascal: Pascal is the force over an area and used as a measurement of pressure. So, 1 Pascal = 1 Newton force spread on 1 square meter area.1 Pascal = 0.00015 PSI

Newton: Newton is the amount of force which helps in accelerating 1 kg mass by 1 meter / second. 1 Newton = 0.225 pounds.

• **Ques: 32. Define 'electric charge'?**

Solution:

Electric charge is the physical property of matter that causes it to experience a force when placed in an electromagnetic field. There are two types of electric charges; positive and negative (commonly carried by protons and electrons respectively).

• **Ques: 33. Define 'power'?**

Solution:

Electrical power is the rate at which electrical energy is converted to another form, such as motion, heat, or an electromagnetic field.

• **Ques: 34. What do you understand by Transducers?**

Solution:

A transducer is a device that converts energy from one form to another. Usually, a transducer converts a signal in one form of energy to a signal in another.

Transducers that convert physical quantities into mechanical ones are called mechanical transducers. Transducers that convert physical quantities into electrical are called electrical transducers.

The examples are a thermocouple that changes temperature differences into a small voltage, or a linear variable differential transformer (LVDT) used to measure displacement.

• **Ques: 35. Define the 'sensor' and 'actuator'?**

Solution:

Sensors and actuators are comprehensive classes of transducers. Some transducers can operate as a sensor or as an actuator, but not as both simultaneously.

A sensor is a transducer that receives and responds to a signal or stimulus from a physical system. It produces a signal, which represents information about the system, which is used by some type of telemetry, information or control system.

Sensor examples: Hotwire anemometers (measure flow velocity), Microphones (measure fluid pressure), Accelerometers (measure the acceleration of a structure), Gas sensors (measure concentration of specific gas or gases), Humidity sensor, Temperature sensors, etc.

An actuator is a device that is responsible for moving or controlling a mechanism or system. It is controlled by a signal from a control system or manual control. It is operated by a source of energy, which can be a mechanical force, electrical current, hydraulic fluid pressure, or pneumatic pressure, and converts that energy into motion. An actuator is a mechanism by which a control system acts upon an environment.

Actuator examples: Motors (which impose a torque), Force heads (which impose a force), Pumps (which impose either a pressure or a fluid velocity).

- **Ques: 36. What is the difference between the active sensor and a passive sensor?**

Solution:

Active sensors require an external power source to operate, which is called an excitation signal. The signal is modulated by the sensor to produce an output signal. For example, a thermistor does not generate an electrical signal, but by passing an electric current through it, its resistance can be measured by detecting variations in the current or voltage across the thermistor.

Passive sensors, in contrast, generate an electric current in response to an external stimulus which serves as the output signal without the need for an additional energy source. Such examples are a photodiode, and a piezoelectric sensor, thermocouple, etc.

- **Ques: 37. What do you mean by hysteresis?**

Solution:

The magnetization of ferromagnetic substances due to a varying magnetic field lags behind the field. This effect is called hysteresis, and the term is used to describe any system in whose response depends not only on its current state but also upon its history.

- **Ques: 38. What is the antenna effect?**

Solution:

Increasing net length can accumulate more changes while manufacturing of the device due to the ionization process. If this net is connected to the gate of the MOSFET it can damage dielectric property of the gate and causing damage to MOSFET.

- **Ques: 39. What is cloning and buffering?**

Solution:

Cloning: It is a method of optimization that decreases the load of the heavily loaded cell by replacing the cell.

Buffering: It is a method of optimization that is used to insert buffer in high fan-out nets to decrease the delay.

- **Ques: 40. Why NAND gate is preferred than NOR?**

Solution:

At transistor level, the mobility of electrons is normally three times that of holes compared to nor and the NAND gate is faster, less leakage.

- **Ques: 41. Difference between 'Electronics' and 'Electrical'?**

Solution:

Electronics deals with the flow of charge (electron) through non-metal conductors (semi-conductors).

Electrical deals with the flow of charge through metal conductors.

Example: Flow of charge through silicon which is not a metal would come under electronics whereas the flow of charge through copper which is a metal would come under electrical.

- **Ques: 42. What do you mean by resistor, capacitor, and inductor?**

Solution:

Resistors: A Resistor is an electrical device that resists the flow of electrical current. It is a passive device used to control, or impede the flow of, electric current in an electric circuit by providing resistance, thereby developing a drop in voltage across the device. The value of a resistor is measured in ohms and represented by the Greek letter capital omega.

Capacitors: In simple words, we can say that a capacitor is a device used to store and release electricity, usually as the result of chemical action. Also referred to as a storage cell, a secondary cell, a condenser or an accumulator. A Leyden Jar was an early example of a capacitor.

Inductors: An inductor is an electrical device (typically a conducting coil) that introduces inductance into a circuit. An inductor is a passive electrical component designed to provide inductance in a circuit. It is a coil of wire wrapped around an iron core. The simplest form of an inductor is made up of a coil of wire. The inductance measured in henrys is proportional to the number of turns of wire, the wire loop diameter and the material or core the wire is wound around. It stores energy in the form of a magnetic field.

- **Ques: 43. What do you mean by semiconductor devices?**

Solution:

A conductor made with semiconducting material. Semiconductors are made up of a substance with electrical properties intermediate between a good conductor and a good insulator. A semiconductor device conducts electricity poorly at room temperature but has increasing conductivity at higher temperatures. Metalloids are usually good semiconductors.

- **Ques: 44. Why Silicon is preferred over germanium?**

Solution:

The knee voltage of germanium is 0.7eV and silicon is 1.1eV because the atomic size of germanium is bigger. Lesser energy is required to free electrons from the outermost orbit.

1. Low Reverse Leakage Current:

The reverse current in silicon flows in order of nano amperes compared to germanium in which the reverse current is in order of microamperes, because of this the accuracy of non-conduction of the Ge diode in reverse bias falls. Whereas Si diode retains its property to a greater extent i.e., it allows the negligible amount of current to flow.

2. Good Temperature Stability:

Temperature stability of silicon is good, it can withstand in temperature range typically from 140C to 180C whereas germanium is much temperature-sensitive only up to 70C.

3. Low Cost:

Silicon is relatively easy and inexpensive to obtain and process, whereas germanium is a rare material that is typically found with copper, lead or silver deposits. Because of its rarity, germanium is more expensive to work with, thus making germanium diodes more difficult to find (and sometimes more expensive) than silicon diodes.

4. High Reverse Break Down Voltage:

The Si diode has a large reverse breakdown voltage of about 70–100V compared to Ge which has the reverse breakdown voltage around 50V.

5. Large Forward Current:

Silicon is much better for high current applications as it has very high forward current in a range of tens of amperes, whereas germanium diodes have very small forward current in a range of microamperes.

• **Ques: 45. What do you mean by the PN Junction diode? Draw its VI characteristics?**

Solution:

A PN-junction diode is formed when a p-type semiconductor is fused to an n-type semiconductor creating a potential barrier voltage across the diode junction.

If a suitable positive voltage (forward bias) is applied between the two ends of the PN junction, it can supply free electrons and holes with the extra energy they require to cross the junction as the width of the depletion layer around the PN junction is decreased.

A diode is a one-way valve for electricity. Diodes allow the flow of electricity in one direction.

Knee voltage is also known as "cut-in-voltage". The minimum amount of voltage required for conducting the diode is known as "knee voltage" or "cut-in-voltage". The forward voltage at which the current through PN junction starts increasing rapidly is known as knee voltage. The V-I characteristics of PN Junction diode is shown as in Figure 5.7.

Biasing conditions:

1. Zero Bias – No external voltage potential is applied to the PN junction diode.
2. Reverse Bias – The voltage potential is connected negative, (−ve) to the P-type material and positive, (+ve) to the N-type material across the diode which has the effect of increasing the PN junction diode's width.
3. Forward Bias – The voltage potential is connected positive, (+ve) to the P-type material and negative, (−ve) to the N-type material across the diode which has the effect of decreasing the PN junction diodes width.

V-I Characteristics:

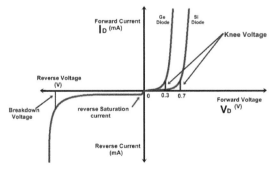

Figure 5.7   V-I characteristics of PN junction diode.

- **Ques: 47. Give the Shockley diode equation?**

Solution:

The Shockley diode equation or the diode law, named after transistor co-inventor William Shockley of Bell Telephone Laboratories, gives the I–V (current-voltage) characteristic of an idealized diode in either forward or reverse bias (applied voltage):

$$I = I_S(e^{\frac{V_D}{nV_T}} - 1)$$

where

I am the diode current,
IS is the reverse bias saturation current (or scale current),
VD is the voltage across the diode,
V_T is the thermal voltage kT/q (Boltzmann constant times temperature divided by electron charge), and n is the ideality factor, also known as the quality factor or sometimes emission coefficient.

The thermal voltage V_T is approximately 25.8563 mV at 300 K (27°C; 80°F). At an arbitrary temperature, it is a known constant defined by:

$$V_T = \frac{kT}{q}$$

where k is the Boltzmann constant, T is the absolute temperature of the p–n junction, and q is the magnitude of the charge of an electron (the elementary charge).

- **Ques: 48. Draw the reverse V-I characteristics of the PN junction diode?**

Solution: The reverse V-I characteristics of PN junction diode is shown in Figure 5.8.

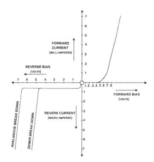

Figure 5.8   Avalanche and Zener breakdown.

Zener Breakdown

When we increase the reverse voltage across the pn junction diode, what happens is that the electric field across the diode junction increases (both internal & external). This results in a force of attraction on the negatively charged electrons at the junction. This force frees electrons from its covalent bond and moves those free electrons to the conduction band. When the electric field increases (with applied voltage), more and more electrons are freed from its covalent bonds. This results in drifting of electrons across the junction and electron-hole recombination occurs. So, a net current is developed and it increases rapidly with an increase in the electric field.

Zener breakdown phenomena occur in a PN junction diode with heavy doping & thin junction (means depletion layer width is very small). Zener breakdown does not result in damage to the diode. Since the current is only due to the drifting of electrons, there is a limit to the increase in current as well.

Avalanche Breakdown

Avalanche breakdown occurs in a pn junction diode which is moderately doped and has a thick junction (means its depletion layer width is high). Avalanche breakdown usually occurs when we apply a high reverse voltage across the diode (obviously higher than the Zener breakdown voltage, say Vz). So as we increase the applied reverse voltage, the electric field across the junction will keep increasing.

If the applied reverse voltage is Va and the depletion layer width is d; then the generated electric field can be calculated as Ea =Va/d

This generated electric field exerts a force on the electrons at the junction and it frees them from covalent bonds. These free electrons will gain acceleration and it will start moving across the junction with high velocity. This results in a collision with other neighboring atoms. These collisions in high velocity will generate further free electrons. These electrons will start drifting and electron-hole pair recombination occurs across the junction. This results in a net current that rapidly increases.

- **Ques: 49. Why avalanche breakdown occurs at a voltage (Va) which is higher than Zener breakdown voltage (Vz)?**

Solution:

The reason behind this is simple. We know, avalanche phenomena occur in a moderately doped diode and junction width (say d) is high. A Zener breakdown occurs in a diode with heavy doping and thin junction (here d is small). The electric field that occurs due to applied reverse voltage (say V) can be calculated as E = V/d.

So in a Zener breakdown, the electric field necessary to break electrons from the covalent bond is achieved with lesser voltage than an avalanche breakdown. The reason is a thin depletion layer width. In avalanche breakdown, the depletion layer width is higher and hence much more reverse voltage has to be applied to develop the same electric field strength (necessary enough to break electrons free)

- **Ques: 50. What do you mean by transistor?**

Solution:

The transistor is a semiconductor device. It is the fundamental building block of the circuitry in mobile phones, computers, and several other electronic devices. A transistor has a very fast response and is used in many functions including voltage regulation, amplification, switching, signal modulation, and oscillators. Transistors may be packaged individually or they can be a part of an integrated circuit. Some of the ICs have billions of transistors in a very small area.

In electronics, a transistor is a semiconductor device commonly used to amplify or switch electronic signals. The transistor is the fundamental building block of computers and all other modern electronic devices. Some transistors are packaged individually but most are found in integrated circuits.

- **Ques: 51. What is a Printed Circuit Board?**

Solution:

A PCB (printed circuit board) or PC board is a piece of phenolic or glass-epoxy board with copper-clad on one or both sides. The portions of copper that are not needed are etched off, leaving "printed" circuits that connect the components.

PCB is used to mechanically support and electrically connect electronic components using conductive pathways, or traces, etched from copper sheets laminated onto a non-conductive substrate.

It is also sometimes referred to as a printed wiring board (PWB) or etched wiring board. A PCB assembled with electronic components is called printed circuit assembly (PCA), OR printed circuit board assembly (PCBA).

• **Ques: 52. What is communication?**

Solution:

Communication means transferring a signal from the transmitter which passes through a medium then the output is obtained at the receiver. Communication is transferring a message from one place to another place.

• **Ques: 53. Different types of communications? Explain.**

Solution:

Analog and digital communication.

As a technology, analog is the process of taking an audio or video signal (the human voice) and translating it into electronic pulses. On the other hand, digital is breaking the signal into a binary format where the audio or video data is represented by a series of "1"s and "0"s.

Digital signals are immune to noise, quality of transmission and reception. Good components used in digital communication can be produced with high precision and power consumption is also very less when compared with analog signals.

• **Ques: 54. What is sampling?**

Solution:

The process of obtaining a set of samples from a continuous function of time x(t) is referred to as sampling.

• **Ques: 55. State sampling theorem.**

Solution:

It states that while taking the samples of a continuous signal, it has to be taken care that the sampling rate is equal to or greater than twice the cut off frequency and the minimum sampling rate is known as the Nyquist rate.

$$Sampling\ rate \geq 2(cut - off\ frequency)$$

The minimum sampling rate is the Nyquist rate.

• **Ques: 56. What is the cut-off frequency?**

Solution:

The frequency at which the response is -3dB concerning the maximum response.

- **Ques: 57. What is passband?**

Solution:

Passband is the range of frequencies or wavelengths that can pass through a filter without being attenuated.

- **Ques:58. What is the stopband?**

Solution:

A stopband is a band of frequencies, between specified limits, in which a circuit, such as a filter or a telephone circuit, does not let signals through, or the attenuation is above the required stopband attenuation level.

- **Ques: 59. Explain RF?**

Solution:

Radiofrequency (RF) is a frequency or rate of oscillation within the range of about 3 Hz to 300 GHz. This range corresponds to the frequency of alternating current electrical signals used to produce and detect radio waves. Since most of this range is beyond the vibration rate that most mechanical systems can respond to, RF usually refers to oscillations in electrical circuits or electromagnetic radiation.

- **Ques: 60. What is the modulation? And where it is utilized?**

Solution:

Modulation is the process of varying some characteristic of a periodic wave with an external signal.

Radio communication superimposes this information bearing signal onto a carrier signal.

These high-frequency carrier signals can be transmitted over the air easily and are capable of traveling long distances.

The characteristics (amplitude, frequency, or phase) of the carrier signal are varied by the information-bearing signal.

Modulation is utilized to send an information-bearing signal over long distances.

- **Ques: 61. What is demodulation?**

Solution:

Demodulation is the act of removing the modulation from an analog signal to get the original baseband signal back. Demodulating is necessary because the receiver system receives a modulated signal with specific characteristics and it needs to turn it to base-band.

- **Ques: 62. Name the modulation techniques.**

Solution:

For analog modulation–AM, SSB, FM, PM and SM Digital modulation–OOK, FSK, ASK, Psk, QAM, MSK, CPM, PPM, TCM, OFDM

- **Ques: 63. Explain AM and FM.**

Solution:

AM-Amplitude modulation is a type of modulation where the amplitude of the carrier signal is varied following the information-bearing signal. FM-Frequency modulation is a type of modulation where the frequency of the carrier signal is varied following the information-bearing signal.

- **Ques: 64. Where do we use AM and FM?**

Solution:

AM is used for video signals for example TV. It ranges from 535 to 1705 kHz. FM is used for audio signals for example radio. It ranges from 88 to 108 MHz.

- **Ques: 65. What is the base station?**

Solution:

The base station is a radio receiver/transmitter that serves as the hub of the local wireless network, and may also be the gateway between a wired network and the wireless network.

- **Ques: 66. How many satellites are required to cover the earth?**

Solution:

3 satellites are required to cover the entire earth, which is placed at 120 degrees to each other. The life span of the satellite is about 15 years.

- **Ques: 67. What is a repeater?**

Solution:

A repeater is an electronic device that receives a signal and retransmits it at a higher level and/or higher power, or onto the other side of obstruction so that the signal can cover longer distances without degradation.

- **Ques: 68. What is an Amplifier?**

Solution:

An electronic device or electrical circuit that is used to boost (amplify) the power, voltage or current of an applied signal.

- **Ques: 69. Example of negative feedback and positive feedback?**

Solution:

The example for −ve feedback is – Amplifiers and for +ve feedback is – Oscillators

- **Ques: 70. What is Oscillator?**

Solution:

An oscillator is a circuit that creates a waveform output from a direct current input. The two main types of oscillators are harmonic and relaxation. The harmonic oscillators have smooth curved waveforms, while relaxation oscillators have waveforms with sharp changes.

- **Ques: 71. What is an Integrated Circuit?**

Solution:

An integrated circuit (IC), also called a microchip, is an electronic circuit etched onto a silicon chip. Their main advantages are low cost, low power, high performance, and very small size.

- **Ques: 72. What is crosstalk?**

Solution:

Crosstalk is a form of interference caused by signals in nearby conductors. The most common example is hearing an unwanted conversation on the telephone. Crosstalk can also occur in radios, televisions, networking equipment, and even electric guitars.

- **Ques: 73. What is the resistor?**

Solution:

A resistor is a two-terminal electronic component that opposes an electric current by producing a voltage drop between its terminals in proportion to the current, that is, following Ohm's law: $V = IR$.

- **Ques: 74. What is an inductor?**

Solution:

An inductor is a passive electrical device employed in electrical circuits for its property of inductance. An inductor can take many forms.

• **Ques: 75. What is a conductor?**

Solution:

A substance, body, or device that readily conducts heat, electricity, sound, etc. Copper is a good conductor of electricity.

• **Ques: 76. What is op-amp?**

Solution:

An operational amplifier often called an op-amp, is a DC-coupled high-gain electronic voltage amplifier with differential inputs[1] and, usually, a single output. Typically, the output of the op-amp is controlled either by negative feedback, which largely determines the magnitude of its output voltage gain or by positive feedback, which facilitates regenerative gain and oscillation.

• **Ques: 77. What is the feedback?**

Solution:

Feedback is a process whereby some proportion of the output signal of a system is passed (fed back) to the input. This is often used to control the dynamic behavior of the system.

• **Ques: 78. Advantages of negative feedback over positive feedback.**

Solution:

Much attention has been given by researchers to negative feedback processes because negative feedback processes lead systems towards equilibrium states. Positive feedback reinforces a given tendency of a system and can lead a system away from equilibrium states, possibly causing quite unexpected results.

• **Ques: 79. What are the Barkhausen criteria?**

Solution:

Barkhausen criteria, without which you will not know which conditions, are to be satisfied with oscillations.

"Oscillations will not be sustained if, at the oscillator frequency, the magnitude of the product of the transfer gain of the amplifier and the magnitude of the feedback factor of the feedback network (the magnitude of the loop gain) are less than unity".

The condition of unity loop gain $-A\beta = 1$ is called the Barkhausen criterion. This condition implies that

A $\beta = 1$ and that the overall phase shift of feedback circuit and amplifier must be zero i.e. phase of $- A\beta$ is zero.

It is a mathematical condition to determine whether a linear electronic circuit will oscillate or not.

- **Ques: 80. What is CDMA, TDMA, FDMA?**

Solution:

Code division multiple access (CDMA) is a channel access method utilized by various radio communication technologies. CDMA employs spread-spectrum technology and a special coding scheme (where each transmitter is assigned a code) to allow multiple users to be multiplexed over the same physical channel. By contrast, time division multiple access (TDMA) divides access by time, while frequency-division multiple access (FDMA) divides it by frequency. An analogy to the problem of multiple access is a room (channel) in which people wish to communicate with each other. To avoid confusion, people could take turns speaking (time division), speak at different pitches (frequency division), or speak in different directions (spatial division). In CDMA, they would speak different languages. People speaking the same language can understand each other, but not other people. Similarly, in radio CDMA, each group of users is given a shared code. Many codes occupy the same channel, but only users associated with a particular code can understand each other.

- **Ques: 81. Explain different types of feedback**

Solution:

Types of feedback:

1. Negative feedback: This tends to reduce output (but in amplifiers, stabilizes, and linearizes operation). Negative feedback feeds part of a system's output, inverted, into the system's input; generally with the result that fluctuations are attenuated. Positive feedback: This tends to increase output.

2. Positive feedback, sometimes referred to as "cumulative causation", is a feedback loop system in which the system responds to perturbation (A perturbation means a system, is an alteration of function, induced by external or internal mechanisms) in the same direction as the perturbation. In contrast, a system that responds to the perturbation in the opposite direction is called a negative feedback system.

3. Bipolar feedback: which can either increase or decrease output.

- **Ques: 82. What are the main divisions of the power system?**

Solution:

The generating system, transmission system, and distribution system

- **Ques: 83. What is the Instrumentation Amplifier (IA) and what are all the advantages?**

Solution:

An instrumentation amplifier is a differential op-amp circuit providing high input impedances with ease of gain adjustment by varying a single resistor.

- **Ques: 84. What is meant by the impedance diagram?**

Solution:

The equivalent circuit of all the components of the power system is drawn and they are interconnected is called impedance diagram.

- **Ques: 85. What is the need for a load flow study?**

Solution:

The load flow study of a power system is essential to decide the best operation existing system and for planning the future expansion of the system. It is also essential for designing the power system.

- **Ques: 86. What is the need for base values?**

Solution:

The components of the power system may operate at different voltage and power levels. It will be convenient for analysis of the power system if the voltage, power, current ratings of the components of the power system is expressed regarding a common value called base value.

- **Ques: 87. What is the difference between resistance and impedance?**

Solution:

The resistance exists in dc circuits since it consists of a zero frequency while impedance exists in ac circuits.

Resistance is a concept used for DC (direct currents) whereas impedance is the AC (alternating current) equivalent.

Resistance is due to electrons in a conductor colliding with the ionic lattice of the conductor meaning that electrical energy is converted into heat. Different materials have different resistivities (a property defining how resistive material of given dimensions will be).

However, when considering AC you must remember that it oscillates like a sine wave so the sign is always changing. This means that other effects need to be considered, namely, inductance and capacitance.

Inductance is most obvious in the coiled wire. When a current flows through a wire a circular magnetic field is created around it. If you coil the wire into a solenoid the fields around the wire sum up and you get a magnetic field similar to that of a bar magnet on the outside but you get a uniform magnetic field on the inside. With AC since the sign is always changing the direction of the field in the wires is always changing - so the magnetic field of the solenoid is also changing all the time. Now when field lines cut across a conductor an emf is generated in such a way to reduce the effects that created it (this is a combination of Lenz's and Faraday's laws which state mathematically that E=N*d(thi)/dt , where thi is the magnetic flux linkage). This means that when an AC flows through a conductor a small back emf or back current is induced reducing the overall current.

Capacitance is a property best illustrated by two metal plates separated by an insulator (which we call a capacitor). When current flows electrons build up on the negative plate. An electric field propagates and repels electrons on the opposite plate making it positively charged. Due to the build-up of electrons on the negative plate, incoming electrons are also repelled so the total current eventually falls to zero in exponential decay. The capacitance is defined as the charge stored/displaced across a capacitor divided by the potential difference across it and can also be calculated by the size of the plates and the primitivity of the insulator.

So simply resistance and impedance have different fundamental origins even though the calculation for their value is the same:

$$R = V/I$$

Impedance is a more general term for resistance that also includes reactance.

In other words, resistance is the opposition to a steady electric current. Pure resistance does not change with frequency, and typically the only time only resistance is considered is with DC (direct current – not changing) electricity.

However, reactance is a measure of the type of opposition to AC electricity due to capacitance or inductance. This opposition varies with frequency. For example, a capacitor only allows DC to flow for a short while until it is charged; at that point, the current will stop flowing and it will look like an open. However, if a very high frequency is put across that capacitor (a signal that has a voltage which is changing very quickly back and forth), the capacitor will look like a short circuit. The capacitor has a reactance which is inversely proportional to frequency. An inductor has a reactance which is directly proportional to frequency DC flows through easily while high-frequency AC is stopped.

Impedance is the total contribution of both resistance and reactance.

• **Ques: 88. What is the difference between resistance and reactance?**

Solution:

Resistance is the measure of opposition to the current flow offered by the material. Usually denoted by R.

Reactance is the resistance offered to the AC currents by inductors and capacitors only. Usually denoted by X.

For capacitors $X = 1/(2\pi fC)$; where f is the frequency, C is the capacitance.

For inductors $X = 2\pi fL$; where f is the frequency,:L is the inductance.

Impedance is the sum of the resistance and reactance of a circuit denoted by $Z = R + jX$ (for primarily inductive circuits) or $Z = R - jX$ (for primarily capacitive circuits).

where $j = \sqrt{(-1)}$.

There are two major differences:

1. Resistance is independent of the frequency of input signal whereas reactance is a quantity that depends on the frequency of the input signal. (Inductive reactance is directly proportional to the frequency and capacitive reactance is inversely proportional to the frequency)
2. Reactance is a wattless quantity. Energy is stored in the form of energy and can be used again and therefore the power loss is 0. Resistance is wattfull quantity. A resistor converts the energy it receives and dissipates it in the form of heat.

• **Ques: 89. What is the difference between reactance and reluctance?**

Solution:

Reluctance is a unit measuring the opposition to the flow of magnetic flux within magnetic materials and is analogous to resistance in electrical circuits. ... Capacitance, C, is measured in Farads and has reactance given by $X = 1/(2\pi fC)$

The reactance is dependent upon frequency.

• **Ques: 90. What is the difference between resistance and inductance?**

Solution:

R – resistance is that property of a material which opposes the flow of current. L – inductance is that property of a material which resists the change in current.

• **Ques: 91. What are the values of resistance, reactance, and impedance of resistor, capacitor, and inductor, in series RLC circuit?**

Solution:

In series-RLC circuit: A resistance, a capacitance and an inductance connected in series across an alternating supply are connected as shown in Figure 5.9.

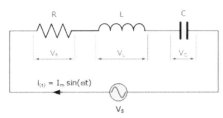

Figure 5.9   Series R-L-C circuit.

In a pure ohmic resistor, the voltage waveforms are "in-phase" with the current. In a pure inductance the voltage waveform "leads" the current by 90°, giving us the expression of ELI. In a pure capacitance the voltage waveform "lags" the current by 90°, giving us the expression of ICE.

This Phase Difference, Φ depends upon the reactive value of the components being used and hopefully by now we know that reactance, (X) is zero if the circuit element is resistive, positive if the circuit element is inductive and negative if it is capacitive thus giving their resulting impedances as:

Table 5.2   Value of R,X and Z in R-L-C circuit

Circuit Element	Resistance (R)	Reactance (X)	Impedance (Z)
Resistor	R	O	$Z_R = R$ $= R\angle 0°$
Inductor	O	ωL	$Z_L = j\omega L$ $= \omega L\angle + 90°$
Capacitor	O	$\dfrac{1}{\omega C}$	$Z_C = \dfrac{1}{j\omega C}$ $= \dfrac{1}{\omega C}\angle - 90°$

The phasor diagram is represented as in Figure 5.10.

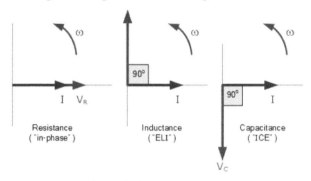

Figure 5.10 Phasor diagram.

- **Ques: 92. Why current leads in the capacitor?**

Solution:

We know for a capacitor that

$$Q = CV$$

where Q is the charge on the capacitor's plates, C is its capacitance, and V is the voltage across the capacitor.

We also know that I, the electric current is the flow of electric charge with time:

$$I = dQ/dt$$

Combine these two, and for a capacitor, we see:

$$I = dQ/dt = C*dV/dt$$

Now, if we have a sinusoidal input voltage, we can calculate the current across the capacitor as a function of the voltage:

$$V(t) = \sin(t)$$
$$I(t) = C*dV(t)/dt = C*\cos(t)$$

But cos(t) is just sin(t) plus pi/2 radians (90 degrees). So our final equations for the capacitor circuit above become:

$$V(t) = \sin(t)$$
$$I(t) = C*\cos(t) = C*\sin(t + pi/2)$$

So, for a sinusoidal input voltage, we see that we also get a sinusoidal current, but the current leads the voltage by pi/2 radians (90 degrees).

• **Ques: 93. Why current lags by voltage by 90 degrees in the inductor?**

Solution:

Inductor: Inductors are storage elements that store energy in the form of magnetic fields. The current flowing through the inductor is the reason for the establishment of the magnetic field.

Lenz's law states that the effect tends to oppose the cause. Try to follow my reasoning here. A voltage was applied across the inductor. This causes current to flow through it. The current passing through the inductor establishes a magnetic field (corkscrew rule). According to Faraday's law, when a current-carrying conductor interacts with a magnetic field, there is an EMF induced in the conductor. So, the current flowing through the inductor and the magnetic field established due to the current interact with each other to induce an EMF in the inductor, which opposes the applied voltage (effect opposes cause) according to Lenz's law. This EMF causes a current to flow, which opposes the main current. This causes a reduction in the main current.

Another way of looking at it is that if you apply your source voltage to an inductor,

$V(S) = \sin(wt)$ then the current must be

$$I = \sin(wt)/R$$

So, by the inductor formula:

$$V(L) = L \, di/dt = L/R \, d \, \sin(wt)/dt = L/R \, \cos(wt)$$

And we know that

$\cos(wt)$ is 90 degrees ahead of $\sin(wt)$.

• **Ques: 94. What is the difference between EMF and voltage?**

Solution:

The EMF is measured between the endpoint of the source, when no current flow through it, whereas, the voltage is measured between any two points of the closed circuit. The EMF is generated by the electrochemical cell, dynamo, photodiodes, etc., whereas the voltage is caused by the electric and magnetic field.

• **Ques: 95. What do you mean by the bathtub curve?**

Solution:

The bathtub curve as shown in Figure 5.11 is widely used in reliability engineering. It describes a particular form of the hazard function which comprises three parts:

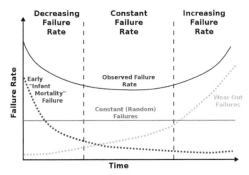

Figure 5.11 Bathtub curve.

The first part is a decreasing failure rate known as early failures.

The second part is a constant failure rate known as random failures.

The third part is an increasing failure rate known as wear-out failures.

The name is derived from the cross-sectional shape of a bathtub: steep sides and a flat bottom.

• **Ques: 96. What do you mean by MTBF, MTTR, and MTTF?**

Solution:

Mean Time Between Failure (MTBF) is a reliability term used to provide the number of failures per million hours for a product. This is the most common inquiry about a product's life span and is important in the decision-making process of the end-user. MTBF is more important for industries and integrators than for consumers. Most consumers are price driven and will not consider MTBF, nor is the data often readily available. On the other hand, when equipment such as media converters or switches must be installed into mission-critical applications, MTBF becomes very important. Also, MTBF may be an expected line item in an RFQ (Request For Quote). Without the proper data, a manufacturer's piece of equipment would be immediately disqualified.

Mean Time To Repair (MTTR) is the time needed to repair a failed hardware module. In an operational system, repair generally means replacing a failed hardware part. Thus, hardware MTTR could be viewed as the mean time to replace a failed hardware module. Taking too long to repair a product drives up the cost of the installation in the long run, due to downtime until the new part arrives and the possible window of time required to schedule the installation. To avoid MTTR, many companies purchase spare products so that a replacement can be installed quickly. Generally, however, customers will inquire about the turn-around time of repairing a product, and indirectly, that can fall into the MTTR category.

Mean Time To Failure (MTTF) is a basic measure of reliability for non-repairable systems. It is the meantime expected until the first failure of a piece of equipment. MTTF is a statistical value and is meant to be the mean over a long period and a large number of units. Technically, MTBF should be used only about a repairable item, while MTTF should be used for non-repairable items. However, MTBF is commonly used for both repairable and non-repairable items.

$$MTBF = MTTR + MTTF$$

- **Ques: 97. What is the significance of FIT?**

Solution:

Failure In Time (FIT) is a way of reporting MTBF. FIT reports the number of expected failures per one billion hours of operation for a device. This term is used particularly by the semiconductor industry but is also used by component manufacturers. FIT can be quantified in many ways: 1000 devices for 1 million hours or 1 million devices for 1000 hours each, and other combinations. FIT and CL (Confidence Limits) are often provided together. In common usage, a claim to 95% confidence in something is normally taken as indicating virtual certainty. In statistics, a claim to 95% confidence simply means that the researcher has seen something occur that only happens one time in twenty or less. For example, component manufacturers will take a small sampling of a component, test x number of hours, and then determine if there were any failures in the testbed. Based on the number of failures that occur, the CL will then be provided as well.

- **Ques: 98. Name any five VLSI companies?**

Solution:
1. ST Microelectronics
2. Synopsys
3. XILINX
4. Cadence Design System
5. Intel

- **Ques: 99. What are the characteristics of an ideal opamp?**

Solution:
1. Infinite open-loop gain
2. Infinite input impedance

3. Zero output impedance
4. Infinite frequency bandwidth
5. The infinite common-mode rejection ratio

• **Ques: 100. What do you mean by CMRR and Slew rate?**

Solution:

CMMR stands for Common Mode Rejection Ratio and it is defined as the ratio of differential voltage gain to the common-mode voltage gain

$$CMMR = Ad/Ac$$

Where Ad is Differential voltage gain and Ac is a common-mode voltage gain

The slew rate of an op-amp or an amplifier circuit is the rate of change in the output voltage caused by a step change in the input. It is measured as a voltage change in a given time – typically V/ṭs or V/ms.

• **Ques: 101. What do you mean by the transient response and steady-state response?**

Solution:

In a system, when certain input changes, it takes a while for the output to stabilize and reach its final state. This interim phase is called the transient phase. The final state is the steady-state and the system will stay there indefinitely until some input changes again.

• **Ques: 102. Draw voltage transfer curve of OpAmp?**

Solution:

This means that the output voltage is directly proportional to the input difference voltage only until it reaches the saturation voltages and thereafter the output voltage remains constant. Thus curve is called an ideal voltage transfer curve, as shown in Figure 5.12 ideal because output offset voltage is assumed to be zero.

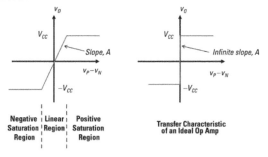

Figure 5.12   Transfer curve of OpAmp.

- **Ques: 103. What do you mean by input offset voltage?**

Solution:

It is the voltage that is applied between the input and output terminal of OpAmp to balance the amplifier and null the output voltage. The input offset voltage is defined as the voltage that must be applied between the two input terminals of the op-amp to obtain zero volts at the output.

- **Ques: 104. Differentiate between microprocessors and microcontrollers?**

Solution:

1. The major difference in both of them is the presence of external peripheral, where microcontrollers have RAM, ROM, EEPROM embedded in it while we have to use external circuits in case of microprocessors.

2. As all the peripheral of the microcontroller is on the single-chip it is compact while the microprocessor is bulky.

3. Microcontrollers are made by using complementary metal-oxide-semiconductor technology so they are far cheaper than microprocessors. Also, the applications made with microcontrollers are cheaper because they need lesser external components, while the overall cost of systems made with microprocessors is high because of the high number of external components required for such systems.

4. The processing speed of microcontrollers is about 8 MHz to 50 MHz, but in contrary processing speed of general microprocessors is above 1 GHz so it works much faster than microcontrollers.

5. Generally, microcontrollers have a power-saving system, like idle mode or power-saving mode so overall it uses less power and also since external components are low overall consumption of power is less. While in microprocessors generally there is no power saving system and also many external components are used with it, so its power consumption is high in comparison with microcontrollers.

6. Microcontrollers are compact so it makes them a favorable and efficient system for small products and applications while microprocessors are bulky so they are preferred for larger applications.

7. Tasks performed by microcontrollers are limited and generally less complex. While tasks performed by microprocessors are software development, Game development, website, document-making, etc. which are generally

more complex so require more memory and speed so that's why external ROM, RAM are used with it.

8. Microcontrollers are based on Harvard architecture where program memory and data memory are separate while microprocessors are based on the von Neumann model where program and data are stored in the same memory module.

Arduino UNO is microprocessor board based on ATmega328P

• **Ques: 105. What do you know about MOSFET?**

Solution:

The MOSFET (as shown in Figure 5.13) is a metal oxide semiconductor field-effect transistor, which is controlled by voltage rather than current. It works electronically, by varying the width of the channel along with the charge carriers flow. Wider the channel, better the device conductance.

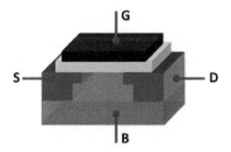

Figure 5.13 MOSFET structure.

The charge carriers enter through the source and exit through the drain. The width of the channel is controlled by the voltage on channel i.e. gate, which is located physically in between the source and drain and insulated from the channel by an extremely thin layer of metal oxide.

• **Ques: 106. Explain how MOSFET functions?**

Solution:

1. Depletion mode:

When there is no voltage on the gate, the channel shows its maximum conductance. As the voltage on the gate is either positive or negative, the channel conductivity decreases.

2. Enhancement mode:

When there is no voltage on the gate the device does not conduct. More is the voltage on the gate, the better the device can conduct.

• **Ques: 107. Draw the characteristics of depletion and enhancement type NMOS and PMOS?**

Solution: The characteristics of NMOS and PMOS is shown as in Figure 5.14.

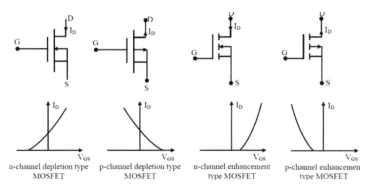

n-channel depletion type MOSFET   p-channel depletion type MOSFET   n-channel enhancement type MOSFET   p-channel enhancemen type MOSFET

Figure 5.14   Types of MOSFET.

• **Ques: 108. Draw I-V characteristics of MOSFET?**

Solution: The I-V characteristics of MOSFET is shown in Figure 5.15.

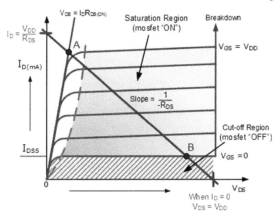

Figure 5.15   V-I characteristics of MOSFET.

• **Ques: 109. What do you understand by pinch-off voltage?**

Solution:

Pinch off voltage: Pinch off voltage is the drain to source voltage after which the drain to source current becomes almost constant and JFET enters into saturation region and is defined only when the gate to source voltage is zero.

• **Ques: 110. Explain MOSFET as a switch?**

Solution: The Figure 5.16 shows MOSFET as a switch.

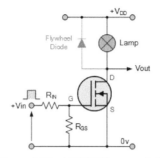

Figure 5.16 MOSFET as a switch.

In this circuit arrangement, an enhanced mode and N-channel MOSFET are being used to switch a sample lamp ON and OFF. The positive gate voltage is applied to the base of the transistor and the lamp is ON (VGS = +v) or at zero voltage level the device turns off (VGS = 0). If the resistive load of the lamp was to be replaced by an inductive load and connected to the relay or diode which is protected to the load. In the above circuit, it is a very simple circuit for switching a resistive load such as a lamp or LED. But when using MOSFET to switch either inductive load or capacitive load protection is required to contain the MOSFET device. We are not giving the protection the MOSFET device is damage. For the MOSFET to operate as an analog switching device, it needs to be switched between its cutoff region where VGS = 0 and saturation region where VGS = +v.

• **Ques: 111. What is forward transconductance in MOSFET?**

Solution:

The forward transconductance is the ratio of Id and (Vgs-Vgs(th)). In the MOSFET switching circuit, it determines the clamping voltage level of gate-source voltage and thus influences the $\frac{d}{dt}V.DS$ during turn-on and turn-off.

• **Ques: 112. What are the three regions of operation in MOSFET?**

Solution:

The three operational regions of MOSFET are:

1. Cut-off Region

With VGS < Vthreshold the gate-source voltage is much lower than the transistors threshold voltage so the MOSFET transistor is switched "fully-OFF" thus, ID = 0, with the transistor acting as an open switch regardless of the value of VDS.

2. Linear (Ohmic) Region

With VGS > Vthreshold and VDS < VGS, the transistor is in its constant resistance region behaving as a voltage-controlled resistance whose resistive value is determined by the gate voltage, VGS level. It is also known as the triode region.

3. Saturation Region

With VGS > Vthreshold and VDS > VGS, the transistor is in its constant current region and is, therefore "fully-ON". The Drain current ID = Maximum with the transistor acting as a closed switch.

- **Ques: 113. Draw I-V characteristics of BJT?**

Solution: The I-V characteristics of BJT is shown in Figure 5.17.

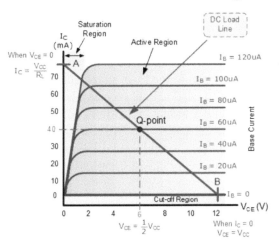

Figure 5.17   V-I characteristics of BJT.

Q-point is an acronym for the quiescent point. Q-point is the operating point of the transistor (ICQ, VCEQ) at which it is biased. The concept of Q-point is used when the transistor act as an amplifying device and hence is operated in the active region of input-output characteristics.

- **Ques: 114. What do you mean by Zener diode?**

Solution:

A Zener diode is a type of diode that allows current to flow not only from its anode to its cathode but also in the reverse direction when the Zener voltage is reached. Zener diodes have a highly doped p–n junction.

The below diagram (Figure 5.18) shows the V-I characteristics of the Zener diode behavior. When the diode is connected in forward-bias diode acts as a normal diode. When the reverse bias voltage is greater than a predetermined voltage then the Zener breakdown voltage occurs.

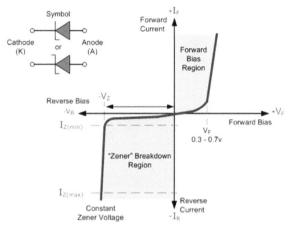

Figure 5.18   Zener diode.

- **Ques: 115. What do you mean by Schmitt trigger?**

Solution:

It is a bi-stable circuit in which the output increases to a steady maximum when the input rises above a certain threshold and decreases almost to zero when the input voltage falls below another threshold, as shown in Figure 5.19.

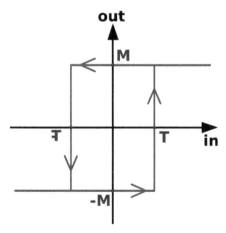

Figure 5.19   Schmitt trigger.

Schmitt trigger is an electronic circuit with positive feedback which holds the output level until the input signal to the comparator is higher than the threshold. It converts a sinusoidal or any analog signal to a digital signal.

• Ques: 116. Draw the CE, CB and CC configuration of BJT?

Solution: The Figure 5.20 shows CE, CB and CC configuration of BJT.

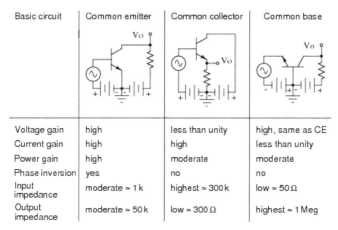

	Basic circuit	Common emitter	Common collector	Common base
Voltage gain		high	less than unity	high, same as CE
Current gain		high	high	less than unity
Power gain		high	moderate	moderate
Phase inversion		yes	no	no
Input impedance		moderate ≈ 1k	highest ≈ 300k	low ≈ 50Ω
Output impedance		moderate ≈ 50k	low ≈ 300 Ω	highest ≈ 1 Meg

Figure 5.20   CE, CB and CC configuration of BJT.

The output characteristics of the CE configuration are shown in Figure 5.21.

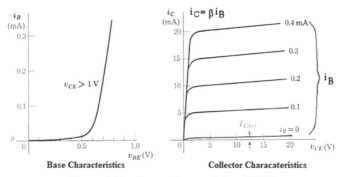

Figure 5.21   Output characteristics of BJT.

• Ques: 117. What is the relation between α and β in the case of BJT?

Solution:

The current gain or current transfer ratio of this CB circuit, denoted by α is defined as the ratio between collector current treated as the output and the emitter current treated as the input:

$$\alpha = \frac{I_C}{I_E} < 1, \quad \text{e.g. } \alpha = 99\% \approx 1$$

The current gain of the CE circuit, denoted by β, is defined as the ratio between the collector current treated as the output and the base current treated as the input:

$$\beta = \frac{I_C}{I_B} = \frac{\alpha I_E}{(1 - \alpha) I_E} = \frac{\alpha}{1 - \alpha},$$

- **Ques: 118. Explain validation and verification in VLSI design?**

Solution:

In Simple terms Verification is Pre-Silicon and Validation is Post Silicon.

Verification is generally used for frontend i.e. the actual verification of the RTL, which can be done mainly using System Verilog/Verilog and using a methodology like OVM, UVM (where we create the entire test environment, which can include transactor, scoreboard, monitor, packets)

The most important thing to note here is to understand that in the Verification step, the input to a verification engineer is Specification and he checks if the RTL designer has coded the given specification accordingly.

Validation terminology is used once the silicon is back from the lab and the team intends to check if the chip is fabricated well and is still functions as it was supposed to before it went for fabrication with no error.

The most important thing to note here is that here the input is not from Specifications i.e. we are not validating the specifications and the chip, but instead, we assume verification engineer has done his part well, we check here if the device is fabricated as per what was given to the fab thus validating the silicon.

- **Ques: 119. What do you understand by black-box testing?**

Solution:

Testing based on an analysis of the specification of a piece of software without reference to its internal workings. The goal is to test how well the component conforms to the published requirements for the component.

- **Ques: 120. What is the difference between a pilot and beta testing?**

Solution:

The difference between a pilot and beta testing is that pilot testing is nothing but using the product (limited to some users) and in beta testing, we do not input real data, but it's installed at the end customer to validate if the product can be used in production.

- **Ques: 121. What's the difference between System Testing and Acceptance testing?**

Solution:

Acceptance testing checks the system against the "Requirements." It is similar to System testing in that the whole system is checked but the important difference is the change in focus. The system testing checks that the system that was specified has been delivered. Acceptance testing checks that the system will deliver what was requested. The customer should always do Acceptance testing and not the developer.

- **Ques: 122. Explain the term 'defect'?**

Solution:

The variation between the actual results and expected results is known as a defect.

If a developer finds an issue and corrects it by himself in the development phase then it's called a defect.

- **Ques: 123. Explain the term 'bug'?**

Solution:

If testers find any mismatch in the application/system in the testing phase then they call it Bug.

- **Ques: 124. Explain the term 'error'?**

Solution:

We can't compile or run a program due to coding mistakes in a program. If a developer unable to successfully compile or run a program then they call it an error.

- **Ques: 125. Explain the term 'failure'?**

Solution:

Once the product is deployed and customers find any issues then they call the product as a failure product. After release, if an end-user finds an issue then that particular issue is called a failure.

- **Ques: 126. What are the different types of 'defect'?**

Solution:

Severity defines the degree of impact. Thus, defect's severity reflects the degree or intensity of a particular defect, to impact a software product or its working, adversely. Based on the severity metric, a defect may be further categorized into the following:

- Critical: The defects termed as 'critical', needs immediate attention and treatment. A critical defect directly affects the critical and essential functionalities, which may affect a software product or its functionality on a large scale, such as failure of a feature/functionality or the whole system, system crash-down, etc.
- Major: Defects, which are responsible for affecting the core and major functionalities of a software product. Although, these defects do not result in complete failure of a system, but may bring several major functions of the software, to rest.
- Minor: These defects produce minor impact, and does not have any significant influence on a software product. The results of these defects may be seen in the product's working; however, it does not stop users to execute a task, which may be carried out, using some other alternative.
- Trivial: These types of defects, have no impact on the working of a product, and sometimes, it is ignored and skipped, such as spelling or grammatical mistakes.
- **Ques: 127. What is the difference between Quality Assurance, Quality Control, and testing?**

Solution:

Quality Assurance is the process of planning and defining the way of monitoring and implementing the quality (test) processes within a team and organization. This method defines and sets the quality standards of the projects.

Quality Control is the process of finding defects and providing suggestions to improve the quality of the software. The methods used by Quality Control are usually established by quality assurance.

It is the primary responsibility of the testing team to implement quality control.

Testing is the process of finding defects/bugs. It validates whether the software built by the development team meets the requirements set by the user and the standards set by the organization.

Here the main focus is on finding bugs and testing teams acts as a quality gatekeeper.

- **Ques: 128. Why power stripes routed in the top metal layers?**

Solution:

The resistivity of top metal layers is less and hence less IR drop is seen in the power distribution network. If power stripes are routed in lower metal layers this will use a good amount of lower routing resources and therefore it can create routing congestion.

- **Ques: 129. What are several factors to improve the propagation delay of a standard cell?**

Solution:

Improve the input transition to the cell under consideration by upsizing the driver.

Reduce the load seen by the cell under consideration, either by placement refinement or buffering.

If allowed increase the drive strength or replace it with LVT (low threshold voltage) cell.

- **Ques: 130. What are the various ways of timing optimization in synthesis tools?**

Solution:

1. Logic optimization: buffer sizing, cell sizing, level adjustment, dummy buffering, etc.
2. Less number of logics between Flip Flops speedup the design.
3. Optimize drive strength of the cell, so it is capable of driving more load and hence reducing the cell delay.
4. Better selection of design ware components (select timing optimized design ware components).
5. Use LVT (Low threshold voltage) and SVT (standard threshold voltage) cells if allowed.

- **Ques: 131. What are the various techniques to resolve congestion/noise?**

Solution:

Routing and placement congestion all depend upon the connectivity in the netlist, a better floor plan can reduce the congestion.

Noise can be reduced by optimizing the overlap of nets in the design.

- **Ques: 132. What is PVT analysis of circuits and how it can be performed?**

Solution:

Process Voltage Temperature variation analysis is important for analyzing model random mismatch of various parameters like length and width of transistors, supply voltage, the thickness of the oxide, temperature and threshold voltage. It can be done using Monte Carlo Analysis using Cadence.

- **Ques: 133. Write the steps for Monte Carlo Analysis?**

Solution: The Figures 5.22–5.24 show the steps for Monte Carlo Analysis.

1. Open the circuit schematic in Cadence Virtuoso.
2. Go to 'File' and click on 'ADEXL'.
3. Click on 'Open Existing' even if you haven't done this analysis before. A new window opens by default, but if you have an existing view, you will not overwrite it

Figure 5.22   Steps for using Cadence Virtuoso.

4. Click on 'Tests' and click on 'Click to add test'. There are a couple of other ways to add a new test:
   Click on 'Create' and then 'Test'.
   Click on the yellow icon in the toolbar.

5. You can pick the whole schematic to run the analysis or one particular component.

6. To add the test, if you have a previously saved analysis click on 'Session' and 'Load state'. To add a new test, click on 'Setup', 'Model libraries' and add the library files required. Usually, libraries are written in scheme language. If you have an academic library, you will have the files in the Spectre folder. You will have to include '/models/all designs.scs' and '/models/design.scs'.

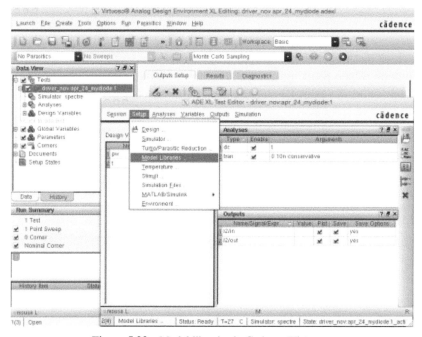

Figure 5.23   Model libraries in Cadence Virtuoso.

7. To add design variables, right-click in the design variable window and click on 'Copy from cell view'.

8. Click on analysis, then choose and select the analysis required. I chose dc and transient analysis.

9. Click on 'Outputs', 'To be plotted', and then 'Select on schematic'. Select the outputs to be plotted.

10. Save the session and go to the ADEXL window again.

11. Click on the green icon beside the drop-down menu saying 'Monte Carlo Sampling'. A window will open as shown in the picture.

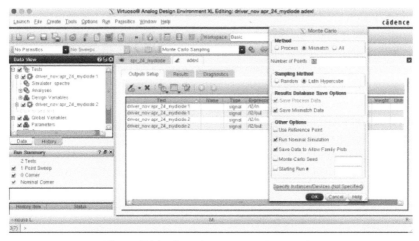

Figure 5.24 Steps for Monte carlo analysis.

12. Enter the number of points, For example, ran the analysis 65 times.
13. Select 'Run nominal simulation' to run a test analysis in addition to the number of runs selected.
14. Click 'OK' to start the simulation.
15. Once the simulation is complete you will see the graph window appear with all the results.

- **Ques: 134. What are the various types of Jobs for VLSI engineers?**

Solution:

1. **Design Engineer:**
   Takes specifications, defines architecture, does circuit design, runs simulations, supervises layout, tapes out the chip to the foundry, evaluate the prototype once the chip comes back from the fab.

2. **Product Engineer:**
   Gets involved in the project during the design phase, ensures manufacturability, develops characterization plan, assembly guidelines, develops quality and reliability plan, evaluates the chip with the design engineer, evaluates the chip through characterization, reliability qualification and manufacturing yield point of view (statistical data analysis). He is responsible for production release and is therefore regarded as a team leader of the project. At post-production, he is responsible for customer returns, failure analysis, and corrective actions including design changes.

3. **Test Engineer:**
   Develops test plan for the chip based on specifications and datasheet, creates characterization and production program for the bench test or the ATE (Automatic Test Equipment), designs test board hardware, correlates ATE results with the bench results to validate silicon to compare with simulation results.

4. **Applications Engineer:**
   Defines new products from a system point of view at the customers' end, based on marketing input. His mission is to ensure the chip works in the system designed or used by the customers and complies with appropriate standards (such as Ethernet, SONET, WiFi, etc.). He is responsible for all customer technical support, firmware development, evaluation boards, datasheets, etc.

5. **Process Engineer:**
   This is a highly specialized function that involves new wafer process development, device modeling, and lots of research and development projects.

6. **Packaging Engineer:**
   This is another highly specialized job function. He develops precision packaging technology, new package designs for the chips, does the characterization of new packages and modeling.

7. **CAD Engineer:**
   This is an engineering function that supports the design engineering function. He is responsible for acquiring, maintaining or developing all CAD tools used by a design engineer. Most companies buy commercially available CAD tools for schematic capture, simulation, synthesis, test vector generation, layout, parametric extraction, power estimation, and timing closure; but in several cases, these tools need some type of customization. A CAD engineer needs to be highly skilled in the use of these tools, be able to write software routines to automate as many functions as possible and have a clear understanding of the entire design flow.

- **Ques: 135. What do you mean by oscillator?**

Solution:

An electronic oscillator is an electronic circuit that produces a periodic, oscillating electronic signal, often a sine wave or a square wave. Oscillators convert direct current (DC) from a power supply to an alternating current (AC) signal. They are widely used in many electronic devices.

- **Ques: 136. What is a ring oscillator?**

Solution:

A ring oscillator (as shown in Figure 5.25) is a device composed of an odd number of NOT gates in a ring, whose output oscillates between two voltage levels, representing true and false. The NOT gates, or inverters, are attached in a chain and the output of the last inverter is fed back into the first.

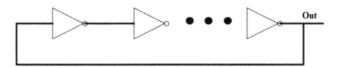

Figure 5.25 Ring oscillator.

The total period of operation is the product of 2*number of gates and gate(inverter) delay. And the frequency of operation will be inverse of period.

It is used as prototype circuits for modeling and designing new semiconductor processes due to simplicity in design and ease of use. It also forms a part of the clock recovery circuit.

- **Ques: 137. What is the basic difference between Analog and Digital Design?**

Solution:

Digital design is distinct from analog design. In analog circuits, we deal with physical signals which are continuous in amplitude and time. Ex: biological data, sensor output, audio, video, etc.

Analog design is quite challenging than digital design as analog circuits are sensitive to noise, operating voltages, loading conditions, and other conditions that have severe effects on performance. Even process technology poses certain topological limitations on the circuit. Analog designer has to deal with real-time continuous signals and even manipulate them effectively even in a harsh environment and in brutal operating conditions.

Digital design, on the other hand, is easier to process and has great immunity to noise. No room for automation in analog design as every application requires a different design. Whereas digital design can be automated. Analog circuits generally deal with the instantaneous value of voltage and current(real-time). It can take any value within the domain of specifications for the device.consists of passive elements that contribute to the noise(thermal) of the circuit. They are usually more sensitive to external noise more so because for a particular function an analog design uses a lot fewer transistors providing design challenges over process corners and temperature ranges. deals with a lot of device-level physics and the state of the transistor plays a very important role Digital Circuits on the other hand deal with only two logic levels 0 and 1(Is it true that according to quantum mechanics there is a third logic level?) deal with lot more transistors for a particular logic, easier to design complex designs, flexible logic synthesis, and greater speed although at the cost of greater power. Less sensitive to noise. design and analysis of such circuits is dependant on the clock. the challenge lies in negating the timing and load delays and ensuring there is no setup or hold violation.

- **Ques: 138. What is stuck at fault?**

Solution:

A Stuck-at fault is a particular fault model used by fault simulators and Automatic test pattern generation (ATPG) tools to mimic a manufacturing defect within an integrated circuit. Individual signals and pins are assumed to be stuck at Logical '1', '0' and 'X'. For example, an output is tied to a logical 1 state during test generation to assure that a manufacturing defect with that type of behavior can be found with a specific test pattern. Likewise, the output could be tied to a logical 0 to model the behavior of a defective circuit that cannot switch its output pin.

- **Ques: 139. What is physical verification?**

Solution:

Physical verification of the design, involves DRC(Design rule check), LVS(Layout versus schematic) Check, XOR Checks, ERC (Electrical Rule Check) and Antenna Checks.

- XOR Check

This step involves comparing two layout databases/GDS by XOR operation of the layout geometries. This check results in a database that has all the mismatching geometries in both the layouts. This check is typically run after a metal spin, wherein the re-spin database/GDS is compared with the previously taped out database/GDS.

- Antenna Check

Antenna checks are used to limit the damage of the thin gate oxide during the manufacturing process due to charge accumulation on the interconnect layers (metal, polysilicon) during certain fabrication steps like Plasma etching, which creates highly ionized matter to etch. The antenna is a metal interconnect, i.e., a conductor like polysilicon or metal, that is not electrically connected to silicon or grounded, during the processing steps of the wafer. If the connection to silicon does not exist, charges may build up on the interconnect to the point at which rapid discharge does take place and permanent physical damage results in the thin transistor gate oxide. This rapid and destructive phenomenon is known as the antenna effect. The Antenna ratio is defined as the ratio between the physical area of the conductors making up the antenna to the total gate oxide area to which the antenna is electrically connected.

• ERC (Electrical rule check)

ERC (Electrical rule check) involves checking a design for all well and substrate areas for proper contacts and spacings thereby ensuring correct power and ground connections. ERC steps can also involve checks for unconnected inputs or shorted outputs.

• **Ques: 140. What are Design Rule Check (DRC) and Layout vs Schematic (LVS)?**

Solution:

Design Rule Check (DRC) and Layout vs Schematic (LVS) are verification processes. Reliable device fabrication at modern deep submicrometre (0.13 ţm and below) requires strict observance of transistor spacing, metal layer thickness, and power density rules. DRC exhaustively compares the physical netlist against a set of "foundry design rules" (from the foundry operator), then flags any observed violations.

Design Rule Check (DRC)

It determines whether the layout of a chip satisfies a series of recommended parameters called design rules. Design rules are a set of parameters provided by semiconductor manufacturers to the designers, to verify the correctness of a mask set. It varies based on the semiconductor manufacturing process. This ruleset describes certain restrictions in geometry and connectivity to ensure that the design has sufficient margin to take care of any variability in the manufacturing process.

Design rule checks are nothing but physical checks of metal width, pitch and spacing requirement for the different layers concerning different manufacturing processes. If we give physical connection to the components without considering the DRC rules, then it will lead to failure of the functionality of the chip, so all DRC violations have to be cleaned up.

After the completion of physical connection, we check every polygon in the design, based on the design rules and report all the violations. This whole process is called the Design Rule Check.

Typical DRC rules are:

- Interior
- Exterior
- Enclosure
- Extension

Layout Versus Schematic (LVS)

LVS is a process that confirms that the layout has the same structure as the associated schematic; this is typically the final step in the layout process. The LVS tool takes as an input a schematic diagram and the extracted view from a layout. It then generates a netlist from each one and compares them. Nodes, ports, and device sizing are all compared. If they are the same, LVS passes and the designer can continue.

LVS tends to consider transistor fingers to be the same as an extra-wide transistor. For example, 4 transistors in parallel (each 1 um wide), a 4-finger 1 um transistor, and a 4 um transistor are all seen as the same by the LVS tool. The functionality of .lib files will be taken from spice models and added as an attribute to the .lib file.

- **Ques: 141. What are the steps involved in semiconductor device fabrication?**

Solution:

This is a list of processing techniques that are employed numerous times in a modern electronic device and do not necessarily imply a specific order.

Wafer processing
Wet cleans
Photolithography
Ion implantation (in which dopants are embedded in the wafer creating regions of increased (or decreased) conductivity)
Dry etching
Wet etching
Plasma ashing
Thermal treatments
Rapid thermal anneal
Furnace anneals

Thermal oxidation
Chemical vapor deposition (CVD)
Physical vapor deposition (PVD)
Molecular beam epitaxy (MBE)
Electrochemical Deposition (ECD). See Electroplating
Chemical-mechanical planarization (CMP)
Wafer testing (where the electrical performance is verified)
Wafer back grinding (to reduce the thickness of the wafer so the resulting chip can be put into a thin device like a smartcard or PCMCIA card.)
Die preparation
Wafer mounting
Die-cutting
IC packaging
Die attachment
IC Bonding
Wire bonding
Flip chip
Tab bonding
IC encapsulation
Baking
Plating
Laser marking
Trim and form
IC testing

- **Ques: 142. Name different types of logic families?**

Solution:

Listed here in rough chronological order of introduction along with their usual abbreviations of Logic family

Diode logic (DL)
Direct-coupled transistor logic (DCTL)
Complementary transistor logic (CTL)
Resistor-transistor logic (RTL)
Resistor-capacitor transistor logic (RCTL)
Diode-transistor logic (DTL)
Emitter coupled logic (ECL) also known as Current-mode logic (CML)
Transistor-transistor logic (TTL) and variants
P-type Metal Oxide Semiconductor logic (PMOS)

N-type Metal Oxide Semiconductor logic (NMOS)
Complementary Metal-Oxide Semiconductor logic (CMOS)
Bipolar Complementary Metal-Oxide Semiconductor logic (BiCMOS)
Integrated Injection Logic (I2L)

- **Ques: 143. What are the different types of IC packaging? Name any 10 types.**

Solution:

IC is packaged in many types they are:

1. BGA1
2. BGA2
3. Ball grid array
4. CPGA
5. The ceramic ball grid array
6. DIP-8
7. Die attachment
8. Dual Flat No-Lead
9. Dual-in-line package
10. Flatpack
11. Land grid array
12. Leadless chip carrier
13. Low insertion force
14. Micro FCBGA
15. Multi-Chip Module
16. Pin grid array
17. Single in-line package
18. Surface-mount technology
19. Through-hole technology
20. Zig-zag in-line package

- **Ques: 144. What is substrate coupling?**

Solution:

In an integrated circuit, a signal can couple from one node to another via the substrate. This phenomenon is referred to as substrate coupling or substrate noise coupling.

The push for reduced cost, more compact circuit boards and added customer features has provided incentives for the inclusion of analog functions on primarily digital MOS integrated circuits (ICs) forming mixed-signal ICs.

• **Ques: 145. What do you understand by 'latchup'?**

Solution:

A latchup is the inadvertent creation of a low-impedance path between the power supply rails of an electronic component, triggering a parasitic structure, which then acts as a short circuit, disrupting the proper functioning of the part and possibly even leading to its destruction due to overcurrent. A power cycle is required to correct this situation. The parasitic structure is usually equivalent to a thyristor (or SCR), a PNPN structure that acts as a PNP and an NPN transistor stacked next to each other. During a latchup when one of the transistors is conducting, the other one begins conducting too. They both keep each other in saturation for as long as the structure is forward-biased and some current flows through it - which usually means until a power-down. The SCR parasitic structure is formed as a part of the totem-pole PMOS and NMOS transistor pair on the output drivers of the gates.

• **Ques: 146. What is a crystal oscillator?**

Solution:

A crystal oscillator (as shown in Figure 5.26) is an electronic oscillator circuit that uses the mechanical resonance of a vibrating crystal of piezoelectric material to create an electrical signal with a precise frequency.

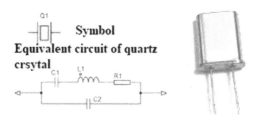

Figure 5.26   Crystal oscillator.

• **Ques: 147. Explain the concept of thyristor?**

Solution:

A thyristor is a solid-state semiconductor device with four layers of alternating P- and N-type materials. It acts exclusively as a bi-stable switch, conducting when the gate receives a current trigger and continuing to conduct until the voltage across the device is reversed biased, or until the voltage is removed.

- **Ques: 148. What is the current-controlled and voltage-controlled device?**

Solution:

Current controlled devices are those, whose output characteristic depends on the input current.

Voltage-controlled devices are those whose output depends on the input voltage.

- **Ques: 149. What do you mean by IGBT?**

Solution:

The insulated gate bipolar transistor (IGBT) is a semiconductor device with three terminals and is used mainly as an electronic switch. It is characterized by fast switching and high efficiency, which makes it a necessary component in modern appliances such as lamp ballasts, electric cars, and variable frequency drives (VFDs).

Its ability to turn on and off, rapidly, makes it applicable in amplifiers to process complex wave-patterns with pulse width modulation. IGBT combines the characteristics of MOSFETs and BJTs to attain high current and low saturation voltage capacity respectively. It integrates an isolated gate using FET (Field effect transistor) to obtain a control input.

- **Ques:150. What is the principle of operation in IGBT?**

Solution:

IGBT requires only a small voltage to maintain conduction in the device unlike in BJT. The IGBT is a unidirectional device, that is, it can only switch ON in the forward direction. This means current flows from the collector to the emitter unlike in MOSFETs, which are bi-directional.

- **Ques: 151. What is cyclo-converter?**

Solution:

A cyclo-converter refers to a frequency changer that can change AC power from one frequency to AC power at another frequency. This process is known as AC-AC conversion. It is mainly used in electric traction, AC motors having variable speed and induction heating.

- **Ques: 152. What do you mean by linear circuit elements?**

Solution:

Linear circuit elements refer to the components in an electrical circuit that exhibit a linear relationship between the current input and the voltage output.

Examples of elements with linear circuits include –

Resistors
Capacitors
Inductors
Transformers

- **Ques: 153. Write the symbol of various resistors?**

Solution:

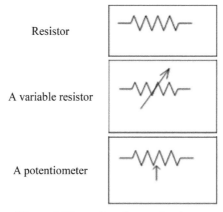

Resistor

A variable resistor

A potentiometer

Figure 5.27 Various forms of resistors.

- **Ques: 154. Write the symbol of various capacitors?**

Solution: The Figure 5.28 shows symbol of various capacitors.

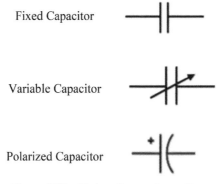

Fixed Capacitor

Variable Capacitor

Polarized Capacitor

Figure 5.28 Various forms of capacitors.

• **Ques: 155. Write the symbol of various inductors?**

Solution: The symbol of various inductors are shown in Figure 5.29.

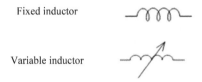

Fixed inductor

Variable inductor

Figure 5.29 Various forms of inductors.

• **Ques: 156. What do you mean by the transformer?**

Solution:

This refers to a device that alters energy from one level to another through a process known as electromagnetic induction. It is usually used to raise or lower AC voltages in applications utilizing electric power.

When the current on the primary side of the transformer is varied, a varied magnetic flux is created on its core, which spreads out to the secondary windings of the transformer in the form of magnetic fields.

The operation principle of a transformer relies on Faraday's law of electromagnetic induction. The law states that the rate of change of the flux linking concerning time is directly related to the EMF induced in a conductor.

A transformer has three main parts (as shown in Figures 5.30 and 5.31) –

• Primary winding
• Magnetic core
• Secondary winding

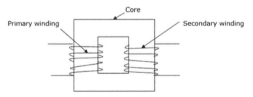

Figure 5.30 Structure of a transformer.

Symbol of a Transformer

Figure 5.31 Symbol of a transformer.

• **Ques: 157. What do you mean by solar cells?**

Solution:

A solar cell or photovoltaic cell is an electrical device that converts the energy of light directly into electricity by the photovoltaic effect, which is a physical and chemical phenomenon. Photovoltaic cells generate electricity by absorbing sunlight and using light energy to create an electrical current. There are many photovoltaic cells within a single solar panel, and the current created by all of the cells together adds up to enough electricity to help power your home.

• **Ques: 158. How many types of solar panels are there?**

Solution:

There are mainly two main types of solar panels:

1. Photovoltaic (PV) solar panels

The technology most people think of when they say "solar panels". These devices convert sunlight into electricity. For the sake of this article, the term "solar panels" will be used to describe photovoltaic panels.

2. Solar thermal collectors

They use the same solar energy that photovoltaic panels do, but they generate heat instead of electrical power.

Multiple solar cells that are oriented in the same way make up what we call solar panels. The electrical power out depends on how many of them are put together.

• **Ques: 159. What are solar cells? Explain its working?**

Solution:

Solar cells (as shown in Figure 5.32) generate an electric current when they are exposed to light. Exactly how this happens is quite complex, and varies between the different types of solar panels.

The basic gist is this:

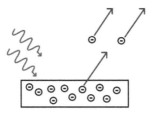

Figure 5.32 A Solar cell.

1. Incoming photons are absorbed by semiconducting material (in most cases silicon) on the surface of the solar cell.
2. These photons "knock loose" electrons from atoms in the solar cell. Since electrons carry a negative charge, an electric potential difference has been created.
3. The solar cell is built in a way that only allows the electron to move in one direction to cancel out the potential.
4. Put many of these reactions together and current starts flowing through the material.

• **Ques: 160. What do you understand by 'LED'?**

Solution:

The lighting emitting diode (Figure 5.33) is a p-n junction diode. It is a specially doped diode and made up of a special type of semiconductor. When the light emits in the forward biased, then it is called as a light-emitting diode.

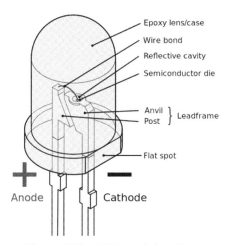

Figure 5.33   A light emitting diode.

• **Ques: 161. What is the working of an LED?**

Solution:

The light-emitting diode simply as shown in Figure 5.34, we know as a diode. When the diode is forward biased, then the electrons & holes are moving fast across the junction and they are combining constantly, removing one another out. Soon after the electrons are moving from the n-type to the p-type silicon, it combines with the holes, then it disappears. Hence it makes the complete atom & more stable and it gives the little burst of energy in the form of a tiny packet or photon of light.

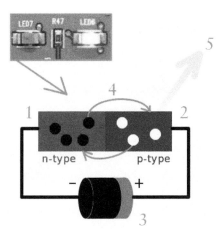

Figure 5.34 Working of LED.

The above diagram shows how the light-emitting diode works and the step by step process of the diagram.

- From the diagram, we can observe that the N-type silicon is in red and it contains the electrons, they are indicated by the black circles.
- The P-type silicon is in blue and it contains holes, they are indicated by the white circles.
- The power supply across the p-n junction makes the diode forward biased and pushing the electrons from n-type to p-type. Pushing the holes in the opposite direction.
- Electron and holes at the junction are combined.
- The photons are given off as the electrons and holes are recombined.

- **Ques: 162. What is the working principle of an LED?**

Solution:

The working principle of the Light-emitting diode is based on quantum theory. The quantum theory says that when the electron comes down from the higher energy level to the lower energy level then, the energy emits from the photon. The photon energy is equal to the energy gap between these two energy levels. If the PN-junction diode is in the forward biased, then the current flows through the diode.

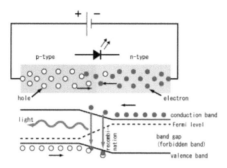

Figure 5.35   Principle of LED.

The flow of current in the semiconductors is caused by both the flow of holes in the opposite direction of the current and flow of electrons in the direction of the current, as shown in Figure 5.35. Hence there will be recombination due to the flow of these charge carriers.

The recombination indicates that the electrons in the conduction band jump down to the valence band. When the electrons jump from one band to another band the electrons will emit the electromagnetic energy in the form of photons and the photon energy is equal to the forbidden energy gap.

For example, let us consider the quantum theory, the energy of the photon is the product of both Planck constant and frequency of electromagnetic radiation. The mathematical equation is shown

$$Eq = hf$$

Where h is known as a Planck constant, and the velocity of electromagnetic radiation is equal to the speed of light i.e c. The frequency radiation is related to the velocity of light as an $f = c / \lambda$. λ is denoted as a wavelength of electromagnetic radiation and the above equation will become as a

$$Eq = hc/\lambda$$

From the above equation, we can say that the wavelength of electromagnetic radiation is inversely proportional to the forbidden gap. In general silicon, germanium semiconductors this forbidden energy gap is between the condition and valence bands are such that the total radiation of electromagnetic waves during recombination is in the form of the infrared radiation. We can't see the wavelength of infrared because they are out of our visible range.

The infrared radiation is said to be as heat because the silicon and the germanium semiconductors are not direct gap semiconductors rather these are indirect gap semiconductors. But in the direct gap semiconductors, the maximum energy level of the valence band and minimum energy level of the conduction band does not occur at the same moment of electrons. Therefore, during the recombination of electrons and holes are migration of electrons from the conduction band to valence band the momentum of the electron band will be changed. The VI characteristics is shown in Figure 5.36.

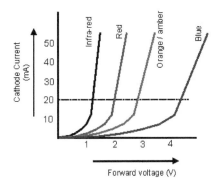

Figure 5.36   Energy level of semiconductor.

• Ques: 163. What do you understand by 'LDR'?

Solution:

A Light Dependent Resistor (LDR), as shown in Figure 5.37 is also called a photoresistor or a cadmium sulfide (CdS) cell. An LDR is a component that has a (variable) resistance that changes with the light intensity that falls upon it. This allows them to be used in light sensing circuits.

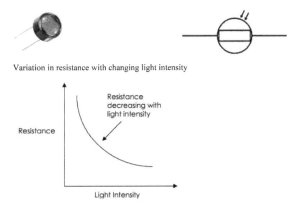

Figure 5.37   A light dependent resistor.

- **Ques: 164. What are the applications of LDR?**

Solution:

There are many applications for Light Dependent Resistors. These include:

1. Lighting switch
   The most obvious application for an LDR is to automatically turn on a light at a certain light level. An example of this could be a street light or a garden light.
2. Camera shutter control
   LDRs can be used to control the shutter speed on a camera. The LDR would be used to measure the light intensity which then adjusts the camera shutter speed to the appropriate level.

- **Ques: 165. Why do we use silicon in solar cells?**

Solution:

Silicon is a semiconductor material. When it is doped with the impurity's gallium and arsenic its ability to capture the sun's energy and convert it into electricity is improved considerably.

- **Ques: 166. What is 'multivibrator'?**

Solution:

It is a device consisting of two amplifying transistors or valves, each with its output connected to the input of the other, which produces an oscillatory signal.

- **Ques: 167. Difference between amplifier and oscillators?**

Solution:

The amplifier is an electronic circuit that gives the output as an amplified form of input. The oscillator is an electronic circuit that gives output without the application of input. The amplifier does not generate any periodic signal.

- **Ques: 168. Why are CE amplifiers widely used?**

Solution:

CE is most widely used because it provides the voltage gain required for most of the day to day applications of preamp and power amps. The common emitter is the most basic configuration for amplifier circuits. It also provides the maximum transconductance or voltage gain for a given load.

- **Ques: 169. Why NPN is preferred over PNP?**

Solution:

Majority charge carriers in NPN are electrons whereas in the case of PNP there are holes. Holes and electrons are charge-carriers in a BJT (NPN or PNP). The difference between them is the mobility (with applied voltage), (effective) mass. Electrons are better when compared to holes so (NPN) is preferred.

- **Ques: 170. Why FET is named field-effect transistor?**

Solution:

FET is named field-effect transistor as field-effect is producing, but in BJT also field effect will produce. The electric field is produced if there is a voltage difference. The field becomes non-zero when voltage differences are non-zero. FET works on the principle that it uses the field produced by the gate in a way that makes the channel conduct more or less.

- **Ques: 171. What is the cut-off frequency?**

Solution:

The frequency at which the response is -3dB concerning the maximum response.

- **Ques: 172. Explain the concept of the rectifier?**

Solution:

A rectifier changes alternating current into direct current. This process is called rectification. The three main types of rectifiers are half-wave, full-wave, and bridge. A rectifier is the opposite of an inverter, which changes direct current into alternating current.

HWR- The simplest type is the half-wave rectifier, which can be made with just one diode. When the voltage of the alternating current is positive, the diode becomes forward-biased and current flows through it. When the voltage is negative, the diode is reverse-biased and the current stops.

The result is a clipped copy of the alternating current waveform with an only positive voltage, and an average voltage that is one-third of the peak input voltage. This pulsating direct current is adequate for some components, but others require a steadier current. This requires a full-wave rectifier that can convert both parts of the cycle to positive voltage.

FWR- The full-wave rectifier is essentially two half-wave rectifiers and can be made with two diodes and an earthed center tap on the transformer. The positive voltage half of the cycle flows through one diode, and the negative half flows through the other. The center tap allows the circuit to be completed because the current cannot flow through the other diode. The result is still a pulsating direct current but with just over half the input peak voltage, and double the frequency.

• **Ques: 173. What are A Transducer and Transponder?**

Solution:

A transducer is a device, usually electrical, electronic, electro-mechanical, electromagnetic, photonic, or photovoltaic that converts one type of energy or physical attribute to another for various purposes including measurement or information transfer.

In telecommunication, the term transponder (short for Transmitter - responder and sometimes abbreviated to XPDR, XPNDR, TPDR or TP) has the following meanings:

- An automatic device that receives, amplifies and retransmits a signal on a different frequency (see also broadcast translator).
- An automatic device that transmits a predetermined message in response to a predefined received signal.
- A receiver-transmitter that will generate a reply signal upon proper electronic interrogation.
- A communications satellite's channels are called transponders because each is a separate transceiver or repeater.

• **Ques: 174. What is an ideal voltage source?**

Solution:

It is a device with zero internal resistance.

• **Ques: 175. What is an ideal current source?**

Solution:

It is a device with infinite internal resistance.

• **Ques: 176. What is a practical voltage source?**

Solution:

It is a device with small internal resistance.

• **Ques: 177. What is a practical current source?**

Solution:

It is a device with large internal resistance.

- **Ques: 178. Design a flow chart depicting various types of capacitors?**

Solution: The Figure 5.38 shows various types of capacitor.

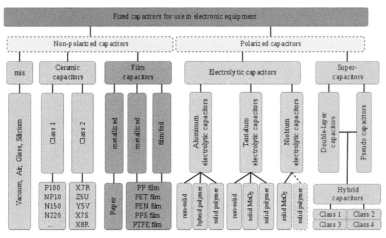

Figure 5.38   Types of capacitors.

- **Ques:179. Show the various capacitors using photographs of components?**

Solution:

Some capacitors are manufactured so they can only tolerate applied voltage in one polarity but not the other. This is due to their construction: the dielectric is a microscopically thin layer of insulation deposited on one of the plates by a DC voltage during manufacture. These are called electrolytic capacitors, and their polarity is marked. The symbols of capacitors are shown in Figure 5.39.

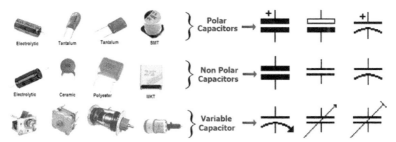

Figure 5.39   Pictorial view of various capacitors.

An electrolytic capacitor (abbreviated e-cap) is a polarized capacitor whose anode or positive plate is made of a metal that forms an insulating oxide layer through anodization. This oxide layer acts as the dielectric of the capacitor.

A polarized ("polar") capacitor (as shown in Figure 5.40) is a type of capacitor that has an implicit polarity - it can only be connected one way in a circuit. The only reason people use polarized caps is that they often cost much less than non-polarized caps of the same capacitance and voltage rating.

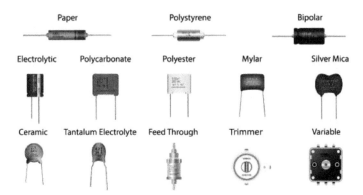

Figure 5.40   Polar capacitors.

- **Ques: 180. Show the various resistors using photographs of components?**

Solution: The picture of various resistors is shown in Figure 5.41.

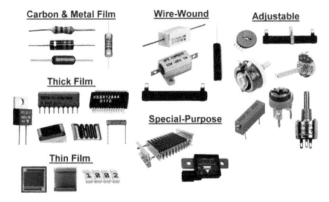

Figure 5.41   Pictorial view of resistors.

- **Ques: 181. Show the various sensors using photographs of components?**

Solution: The pictorial view of various sensors is shown in Figure 5.42.

Figure 5.42 Pictorial view of various sensors.

- **Ques: 182. Classify various types of solar cells?**

Solution: The Figure 5.43 shows the classification of photo-volatic cell.

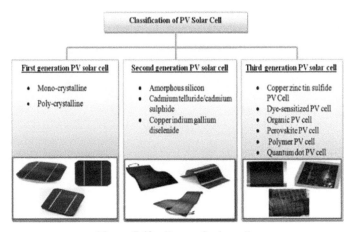

Figure 5.43 Types of solar cells.

• Ques: 183. What causes capacitors to leak?

Solution:

There are two kinds of "leakage" in capacitors.

1. The gradual loss of charge in the capacitor. This occurs because the dielectric material is not a perfect insulator–it has a small but nonzero electrical conductivity, so it acts as a large resistance in parallel with the capacitor. Leakage can also occur through external components connected to the capacitor. This kind of leakage is why dynamic random-access memory (DRAM) must be constantly refreshed.

2. When the capacitor physically leaks its liquid electrolyte. This can occur in a very dramatic way if an electrolytic capacitor is connected backward, or if an internal short circuit develops due to a manufacturing defect. The high current through the short circuit generates gas, which causes pressure to build up. (I'm not sure if the gas is generated through evaporation or electrolysis of the electrolyte, or a combination of both.) This opens a safety vent (a part of the capacitor's casing designed to fail under high pressure), venting the electrolyte. The safety vent is supposed to prevent the pressure from building up to the point where the entire capacitor violently explodes; in cheap capacitors, sometimes it fails.

Capacitors can also leak electrolytes non-violently, for example, due to corrosion of the case.

• Ques: 184. Show the various inductors using photographs of components?

Solution: The Figure 5.44 shows pictorial views of inductor.

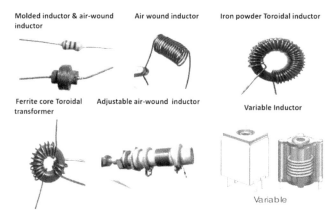

Figure 5.44   Pictorial views of inductors.

• **Ques: 185. Show the symbol of various types of inductors?**

Solution: The various types of inductors is shown in Figure 5.45.

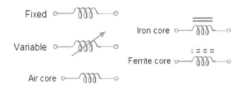

Figure 5.45 Various types of inductors.

Series and Parallel Connections (as shown in Figure 5.46):

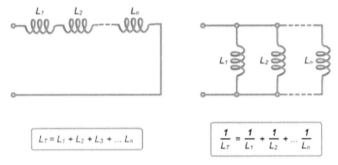

Figure 5.46 Series and parallel configuration.

• **Ques: 186. Show the various optoelectronic devices using photographs?**

Solution: The pictorial view of various optoelectronic devices is shown in Figure 5.47.

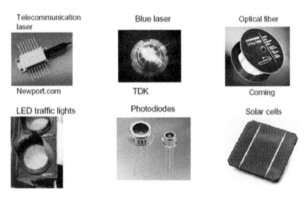

Figure 5.47 Various types of optoelectronic devices.

• **Ques: 187. Explain the 'Lenz law'?**

Solution:

When the current through a coil changes, a voltage is induced. Lenz's Law states that the polarity of the induced voltage always opposes the change in current that caused it, as shown in Figure 5.48. The diagram above illustrates this law. When the switch closes, the current tries to increase, and the magnetic field starts expanding. The expanding magnetic field induces a voltage, which opposes any increase in current. So, at the instant of switching, the current remains the same. When the rate of expansion decreases, the induced voltage decreases, allowing the current to increase. As the current reaches a constant value, there is no induced voltage.

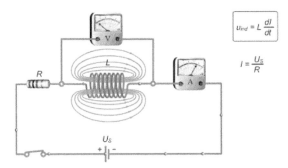

Figure 5.48   Lenz law.

• **Ques: 188. What do you mean by 'sine wave'?**

Solution:

The sine wave (sinusoidal wave or, simply, sinusoid) is the fundamental form of alternating current (AC) and voltage (as shown in Figure 5.49). The current reverses polarity over time. In one cycle, the polarity changes once. The time required for a given sine wave to complete one full cycle is called a period.

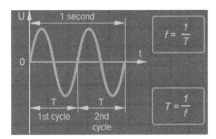

Figure 5.49   Sine wave.

The number of cycles per second is the frequency (f), whose unit is the Hertz (Hz). One hertz is equal to one cycle per second. The frequency and period are reciprocal. More cycles per second results in a higher frequency and a shorter period.

• **Ques: 189. What is SOA for BJT?**

Solution:

The safe operating area (SOA) defines the current and voltage limitations of power devices.

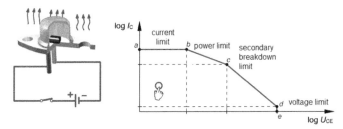

Figure 5.50   Safe operating area.

The above Figure 5.50 shows the typical SOA of a power bipolar transistor. It can be partitioned into four regions. The maximum current limit (section a-b) and the maximum voltage limit (d-e) are determined by the technological features and construction of the particular device. The maximum power dissipation limits the product of the transistor's currents and voltages (section b-c). A secondary breakdown (c-d) occurs when high voltages and high currents appear simultaneously when the device is turned off. When this happens, a hot spot is formed and the device fails due to thermal runaway.

• **Ques: 190. What is SOA for MOSFET & IGBT?**

Solution:

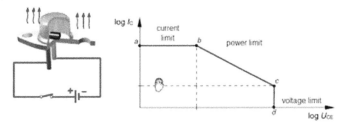

Figure 5.51   SOA for MOSFET and IGBT.

The above Figure 5.51 shows the safe operating area (SOA) of MOSFET transistors. This area is bounded by three limits: current limit (section a–b), maximum power dissipation limit (b–c), and the voltage limits (c–d). The SOA of an IGBT is identical to that of the MOSFET SOA.

Since the drain current decreases when the temperature increases in MOSFET transistors, the possibility of secondary breakdown is almost nonexistent. If local heating occurs, the drain current – and consequently the power dissipation – both diminish. This avoids the creation of local hot spots that can cause thermal runaway.

The above Figure 5.51 demonstrates how the SOA of a device increases when the device is operating in pulse mode. When the device is operating in DC mode the safe operating area is at its smallest. The SOA grows when pulse mode is used. The shorter the pulse signal, the higher the SOA.

- **Ques: 191. What do you understand by maximum power dissipation?**

Solution:

The high currents and voltages in power devices produce very high internal power loss. This loss occurs in the form of heat that must be dissipated; otherwise, the device can be destroyed as a result of overheating.

The maximum power dissipation Pmax indicates a device's maximum capability to transfer and conduct this power loss without overheating.

- **Ques: 192. How maximum power dissipation and temperature are related?**

Solution:

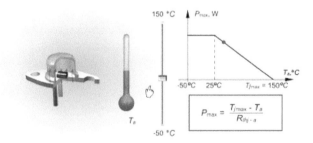

Figure 5.52 Maximum power dissipation.

The maximum power dissipation Pmax of the transistor depends on the highest junction temperature that will not destroy the device Tjmax, the ambient temperature Ta, and the thermal resistance Rthj-a according to the equation shown in the above Figure 5.52.

If the ambient temperature is less than or equal to 25°C the device reaches its maximum specified power rating. When the ambient temperature increases, the power rating decreases. If the ambient temperature Ta reaches the maximum junction temperature Tjmax, maximum power Pmax becomes zero.

• **Ques: 193. What do you mean by 'heat sink'?**

Solution:

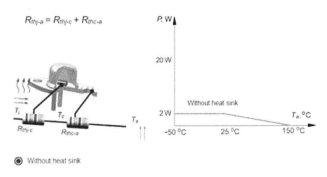

Figure 5.53   Heat sink.

One way to increase the power rating of the device is to diminish the thermal resistance (Rthj-a). A heat sink, which is usually a metal construction with a large surface area, is used to allow heat to dissipate to ambient more easily.

When a heat sink is present, the global thermal resistance (Rthj-a) decreases because there are more paths available for heat dissipation. The case-to-heat-sink thermal resistance (Rthc-s) and the heat-sink-to-ambient thermal resistance (Rths-a) both facilitate heat dissipation. As a result, the power rating increases as illustrated in the above Figure 5.53.

• **Ques: 194. What is the power transistor?**

Solution:

Power Transistors are electronic components that are used for the control and regulation of voltages and currents with high values. They are the basic components for the implementation of linear and switched-mode power supplies, motor control circuits, automotive and aerospace systems, home appliances, and energy management systems

• **Ques: 195. Why power transistors are necessary?**

Solution:

Power Transistors are used to produce, convert, control and regulate high amounts of power output. Typical headphone amplifiers have a low output value (just a fraction of a watt). They are usually implemented with standard low power transistors. On the other hand, amplifiers with a hundred-watt output power are used to ensure quality sound in large rooms or concert halls. These amplifiers operate with high-level currents and voltages (more than dozens of amperes and volts). The output stages of such amplifiers can be implemented only with power transistors.

Power Transistors are capable of providing high currents and high blocking voltages and therefore, high power. They can be classified into BJT (Bipolar Junction Transistors), MOSFET (Metal Oxide Semiconductor Field Effect Transistors), and devices such as IGBT (Insulated Gate Bipolar Transistors) that combine bipolar and MOS technologies.

The principle of operation behind high power transistors is conceptually the same as bipolar or MOS transistors. The main difference is that the active area of the power devices is distinctly higher, resulting in a much higher current handling capacity. For this reason, they have large packages.

• **Ques: 196. Draw NPN BJT, N-channel MOSFET and IGBT device?**

Solution:

The below mentioned Figure 5.54 depicts the typical structure of BJT, MOS-FET, and IGBT devices. For a BJT to maintain conduction, a high continuous current through the base region is required. This imposes the necessity of high-power drive circuits.

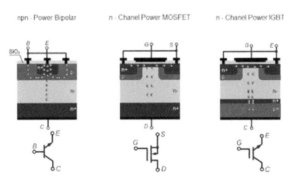

Figure 5.54   Structure of BJT, MOSFET and IGBT.

MOSFETs and IGBTs are voltage-controlled devices. The IGBT has one more junction than the MOSFET, which allows for a higher blocking voltage but limits the switching frequency. In IGBTs, during conduction, the holes from the collector p+ region are injected into the n- region. The accumulated charge reduces IGBT's on-resistance and thus the collector-to-emitter voltage drop is also reduced.

- **Ques: 197. What is the instantaneous and peak value of the sine wave?**

Solution:

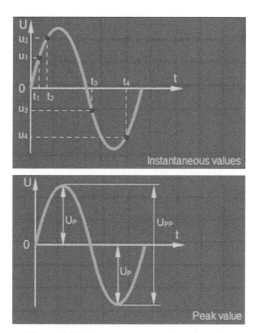

Figure 5.55 Instantaneous and peak value.

At any point in time on a sine wave, the voltage has an instantaneous value. As a cycle represents a continuous set of instantaneous values, other dimensions have been defined to enable comparing one wave to another. The peak value Up is the maximum value. It applies to either the positive or negative peak. The peak-to-peak value, Upp, is the voltage (or current) from the positive peak to the negative peak. The average value is an arithmetic average of all the values in a sine wave for one half-cycle, where Uavr = 0.637 Up

• **Ques: 198. What do you understand by the 'RMS value' of Sine wave?**

Solution:

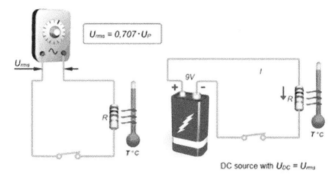

Figure 5.56   RMS value of sine wave.

To compare AC and DC voltage, the effective value of the AC voltage should be calculated using the root-mean-square (RMS) value of the sinusoidal voltage. The (RMS) value of a sinusoidal voltage or current is equal to the dc voltage and current that produces the same heating effect. The formula is Urms = 0.707 Up. The factor 0.707 for RMS value is derived as the square root of the average (mean) of all the squares of the sine wave. To convert from RMS to peak value, the formula Up = 1.414 Urms is used. Unless indicated otherwise, all sine wave ac measurements are in RMS values.

• **Ques: 199. What is the phase angle of the sine wave?**

Solution:

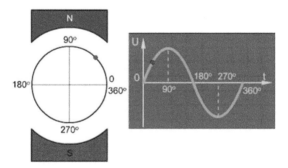

Figure 5.57   Phase angle of sine wave.

The angular measurement of a sine wave can be related to the angular rotation of an AC generator, as shown in the diagram above. It is based on

360o of rotation for the complete cycle of a sine wave. The diagram shows angles in degrees over the full cycle of a sine wave. Since $360° = 2\pi$ rad, angles can be also expressed in radians using the formula in the illustration above.

The phase angle of a sine wave specifies the position of that sine wave relative to a reference. The illustration shows the phase shifts of a sine wave. There is a phase angle of 30o between sine wave A and sine wave B.

- **Ques: 200. What are the Laws of Resistive AC Circuits?**

Solution:

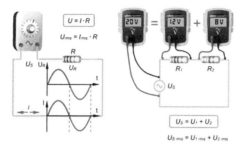

Figure 5.58   Ohm's law and Kirchoff's law.

Ohm's Law and Kirchoff's Law apply to AC circuits in the same way they apply to DC circuits. If a sinusoidal voltage is applied across a resistor, there is a sinusoidal current. It is zero when the voltage is zero and is max. when the voltage is max. The voltage and the currents are in phase with each other.

In a resistive circuit that has an AC voltage source, the source voltage is the sum of all the voltage drops, just as in a DC circuit. Remember, both the voltage and the current must be expressed in the same way, i.e., both in RMS, both in peak, etc.

- **Ques: 201. What are Non-Sinusoidal AC Waveforms?**

Solution:

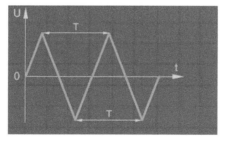

Figure 5.59   Non-sinusoidal AC waveforms.

The pulse and the triangular waveform are the other two major types of signals widely used in electronics. Any waveform that repeats itself at fixed intervals is periodic. The period is denoted with a T. Triangular waveform are formed by voltage or current ramps. A ramp is a linear increase or decrease in the voltage.

An ideal pulse consists of two equal but opposite steps separated by an interval of time called the pulse width. The duty cycle of the pulse is the ratio of the pulse width to the period and is usually expressed in a percentage.

• **Ques: 202. What do you know about breadboard?**

Solution:

A breadboard is a solderless device for a temporary prototype with electronics and test circuit designs. Most electronic components in electronic circuits can be interconnected by inserting their leads or terminals into the holes and then making connections through wires where appropriate.

• **Ques: 203. How series and parallel connections are done using breadboard?**

Solution:

Series connection:

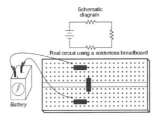

Figure 5.60 Series connection on breadboard.

Parallel Connection:

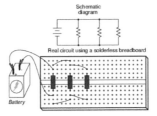

Figure 5.61 Parallel connection on breadboard.

• **Ques: 204. What do you mean by worst-case analysis?**

Solution:

Worst-case circuit analysis (WCCA or WCA) is a cost-effective means of screening design to ensure with a high degree of confidence that potential defects and deficiencies are identified and eliminated before and during the test, production, and delivery.

It is a quantitative assessment of the equipment performance, accounting for manufacturing, environmental and aging effects. In addition to circuit analysis, a WCCA often includes stress and derating analysis, failure modes and effects criticality (FMECA) and reliability prediction (MTBF).

• **Ques: 205. What is the need for design verification and reliability?**

Solution:

1. Verifies circuit operation and quantifies the operating margins over part tolerances and operating conditions
2. Improve circuit performance – Determines the sensitivity of components to certain characteristics or tolerances to better optimize/understand design and what drives performance
3. Verifies that a circuit interfaces with another design properly
4. Determines the impact of part failures or out of tolerance modes.

• **Ques: 206. What is a Catastrophic failure?**

Solution:

A catastrophic failure is a sudden and total failure from which recovery is impossible. Catastrophic failures often lead to cascading systems failure.

• **Ques: 207. How sine waves are produced?**

Solution:

Sine waves are produced electro-magnetically by an AC generator or electronically by an oscillator circuit, which is used in a signal generator.

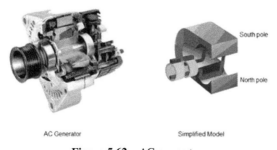

Figure 5.62   AC generator.

The Figure 5.62 shows a cross-section of an AC generator. A simplified model of this generator consists of a single loop of wire in a permanent magnetic field. Magnetic flux lines exist around the north and south poles of the magnet. When a conductor rotates through the magnetic field, a voltage is induced.

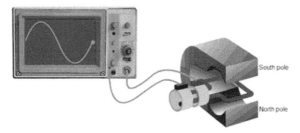

Figure 5.63   Waveform generation.

In a horizontal starting position, the loop does not induce a voltage because the conductors are not cutting across the magnetic flux lines. As the loop rotates through the first quarter of the cycle, it cuts through the flux lines producing the maximum induced voltage. During the second quarter of the cycle, the voltage decreases from its positive maximum back to zero. During the second half of the revolution, the wire loop cuts through the magnetic field in the opposite direction. Thus, the induced voltage has the opposite polarity. After one complete revolution of the loop, one full cycle of the sinusoidal voltage has been completed.

- **Ques: 208. What is a signal generator and Cathode Ray Oscilloscope?**

Solution:

A signal generator (as shown in Figure 5.64) is an instrument that electronically produces sinusoidal voltages or other types of waveforms whose amplitude and frequency can be adjusted. A typical signal generator is shown in the illustration.

Figure 5.64   A signal generator.

An oscilloscope previously called an oscillograph, and informally known as a scope or o-scope, CRO, or DSO is a type of electronic test instrument that graphically displays varying signal voltages, usually as a two-dimensional plot of one or more signals as a function of time.

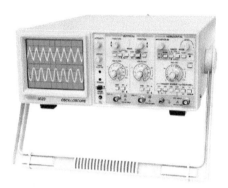

Figure 5.65   A cathode ray oscilloscope.

The cathode ray oscilloscope (as shown in Figure 5.65) is an electronic test instrument, it is used to obtain waveforms when the different input signals are given.

- **Ques: 209. What is the difference between the bypass capacitor and the decoupling capacitor?**

Solution:

The coupling capacitor (as shown in Figure 5.66) is used to maintain the DC biasing condition of circuit (different amplifiers, etc) cannot pass DC through it. So that DC voltage at different terminals remains the same. Whereas, the by-pass capacitor is used to pass the AC which passes through the resistor so that the effective gain will be higher for the same circuit.

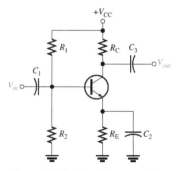

Figure 5.66   Coupling capacitor.

Here C1, C3 are coupling capacitor and C2 is a by-pass capacitor

The terms "bypass capacitor" and "decoupling capacitor" are used interchangeably, though there are definite differences between them.

The bypass capacitor ("bypass") helps us meet this requirement by constraining the unwanted communications a.k.a. the "noise" emanating from the power line to the electronic circuit in question. Any glitch or noise appearing on the power line is immediately bypassed into the chassis ground ("GND") and thus prevented from entering into the system, hence the name bypass capacitor.

Decoupling capacitors ("decap"), on the other hand, are used to isolate two stages of a circuit so that these two stages don't have any DC effect on each other. In reality, decoupling is a refined version of bypassing.

• **Ques: 210. What do you mean by SOC?**

Solution:

As the name suggests, it means shrinking the whole system onto a single chip. The most important feature of the chip is that its functionality should be comparable to that of the original system. It improves quality, productivity, and performance. The system on Chip (SOC) is shown in Figure 5.67.

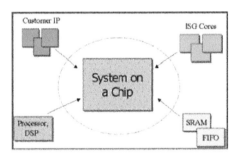

Figure 5.67   System-on-chip.

• **Ques: 211. What do you mean by diode?**

Solution:

A diode is defined as a two-terminal electronic component that only conducts current in one direction. A diode will have negligible resistance in one direction (to allow current flow), and very high resistance in the reverse direction (to prevent current flow).

A PN junction (as shown in Figure 5.68) is the simplest form of the semiconductor diode. In ideal conditions, this PN junction behaves as a short circuit when it is forward biassed, and as an open circuit when it is in the reverse biased. The name diode is derived from "di-ode" which means a device that has two electrodes.

Anode (+) **Cathode (-)**

Figure 5.68 P-N junction diode.

The arrowhead points in the direction of conventional current flow in the forward biased condition. That means the anode is connected to the p side and the cathode is connected to the n side.

- **Ques: 212. What are the main applications of diode?**

Solution:

Diodes can be used as rectifiers, signal limiters, voltage regulators, switches, signal modulators, signal mixers, signal demodulators, and oscillators.

- **Ques: 213. What is the basic principle of the diode?**

Solution:

Semiconductor diodes are the most common type of diode. These diodes begin conducting electricity only if a certain threshold voltage is present in the forward direction (i.e. the "low resistance" direction). The diode is said to be "forward biased" when conducting current in this direction. When connected within a circuit in the reverse direction (i.e. the "high resistance" direction), the diode is said to be "reverse biased".

- **Ques: 214. Why do we say that diodes have a high resistance in the reverse direction, not an infinite resistance?**

Solution:

A diode only blocks current in the reverse direction (i.e. when it is reverse biased) while the reverse voltage is within a specified range. Above this range, the reverse barrier breaks. The voltage at which this breakdown occurs is called the "reverse breakdown voltage". When the voltage of the circuit is higher than the reverse breakdown voltage, the diode can conduct electricity in the reverse direction (i.e. the "high resistance" direction). This is the reason that diodes have a high resistance in the reverse direction, not an infinite resistance.

- **Ques: 215. Give the working principle of PN junction diode in forward bias?**

Solution:

In forward bias, as shown in Figure 5.69 the positive terminal of a source is connected to the p-type side and the negative terminal of the source is connected to the n-type side of the diode and if we increase the voltage of this source slowly from zero, the diode will be in forwarding biased state.

In the beginning, there is no current flowing through the diode. This is because although there is an external electrical field applied across the diode still the majority charge carriers do not get sufficient influence of the external field to cross the depletion region. As we told that the depletion region acts as a potential barrier against the majority charge carriers. This potential barrier is called forward potential barrier. The majority charge carriers start crossing the forward potential barrier only when the value of externally applied voltage across the junction is more than the potential of the forward barrier. For silicon diodes, the forward barrier potential is 0.7 volt and for germanium diodes, it is 0.3 volt.

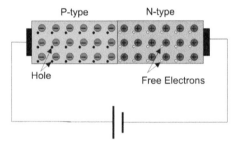

Figure 5.69   Biasing of PN junction diode.

When the externally applied forward voltage across the diode becomes more than the forward barrier potential, the free majority charge carriers start crossing the barrier and contribute the forward diode current. In that situation, the diode would behave as a short-circuited path and the forward current gets limited by only externally connected resistors to the diode.

- **Ques: 216. Give the working principle of PN junction diode in reverse bias?**

Solution:

A diode is said to be operated in reverse bias condition if we connect the negative terminal of the voltage source to the p-type side and positive terminal of the voltage source to the n-type side of the diode. At that condition, due to electrostatic attraction of the negative potential of the source, the holes in the p-type region would be shifted more away from the junction leaving more uncovered negative ions at the junction. In the same way, the free electrons in the n-type region would be shifted more away from the junction towards the positive terminal of the voltage source leaving more uncovered positive ions in the junction. As a result of this phenomenon, the depletion region becomes wider. This condition of a diode is called the reverse biased condition, as shown in Figure 5.70. At that condition, no majority carriers across the junction as they go away from the junction. In this way, a diode blocks the flow of current when it is reverse biased.

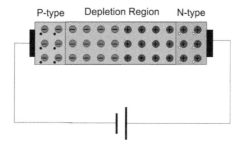

Figure 5.70   Principle of PN junction diode.

There are always some free electrons in the p-type semiconductor and some holes in the n-type semiconductor. These opposite charge carriers in a semiconductor are called minority charge carriers. In the reverse biased condition, the holes find themselves in the n-type side would easily cross the reverse-biased depletion region as the field across the depletion region does not present rather it helps minority charge carriers to cross the depletion region. As a result, there is a tiny current flowing through the diode from positive to the negative side. The amplitude of this current is very small as the number of minority charge carriers in the diode is very small. This current is called reverse saturation current.

If the reverse voltage across a diode gets increased beyond a safe value, due to higher electrostatic force and due to higher kinetic energy of minority charge carriers colliding with atoms, several covalent bonds get broken to contribute a huge number of free electron-hole pairs in the diode and the process is cumulative. The huge number of such generated charge carriers would contribute a huge reverse current in the diode. If this current is not limited by an external resistance connected to the diode circuit, the diode may permanently be destroyed.

- **Ques: 217. Name various types of diodes?**

Solution:

- Zener diode
- P-N junction diode
- Tunnel diode
- Varactor diode
- Schottky diode
- Photodiode
- PIN diode
- Laser diode
- Avalanche diode
- Light emitting diode

- **Ques: 218. Explain the concept of Zener diode in brief?**

Solution:

Zener Diode, as shown in Figure 5.71 is a p-n junction diode connected in reverse bias. But ordinary PN junction diode connected in reverse biased condition is not used as Zener diode practically. A Zener diode is a specially designed, highly doped PN junction diode.

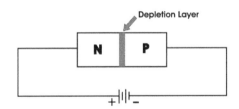

Figure 5.71   Reverse biased PN junction diode.

When a PN junction diode is reverse biased, the depletion layer becomes wider. If this reverse-biased voltage across the diode is increased continually, the depletion layer becomes more and wider. At the same time, there will be a constant reverse saturation current due to minority carriers.

After a certain reverse voltage across the junction, the minority carriers get sufficient kinetic energy due to the strong electric field. Free electrons with sufficient kinetic energy collide with stationary ions of the depletion layer and knock out more free electrons. These newly created free electrons also get sufficient kinetic energy due to the same electric field, and they create more free electrons by collision cumulatively. Due to this commutative phenomenon, very soon, huge free electrons get created in the depletion layer, and the entire diode will become conductive. This type of breakdown of the depletion layer is known as an avalanche breakdown, but this breakdown is not quite sharp.

There is another type of breakdown in the depletion layer which is sharper compared to avalanche breakdown, and this is called Zener breakdown. When a PN junction is diode is highly doped, the concentration of impurity atoms will be high in the crystal. This higher concentration of impurity atoms causes a higher concentration of ions in the depletion layer hence for the same applied reverse-biased voltage, the width of the depletion layer becomes thinner than that in a normally doped diode.

Due to this thinner depletion layer, voltage gradient or electric field strength across the depletion layer is quite high. If the reverse voltage is continued to increase, after a certain applied voltage, the electrons from the covalent bonds within the depletion region come out and make the depletion region conductive. This breakdown is called Zener breakdown. The voltage at which this breakdown occurs is called Zener voltage.

If the applied reverse voltage across the diode is more than Zener voltage, the diode provides a conductive path to the current through it hence, there is no chance of further avalanche breakdown in it. Theoretically, Zener breakdown occurs at a lower voltage level then avalanche breakdown in a diode, especially doped for Zener breakdown. The Zener breakdown is much sharper than the avalanche breakdown. The Zener voltage of the diode gets adjusted during manufacturing with the help of required and proper doping. When a Zener diode is connected across a voltage source, and the source voltage is more than Zener voltage, the voltage across a Zener diode remains fixed irrespective of the source voltage. Although at that condition current through the diode can be of any value depending on the load connected with the diode. That is why we use a Zener diode mainly for controlling voltage in different circuits. The VI characteristics of zener diode is shown in Figure 5.72.

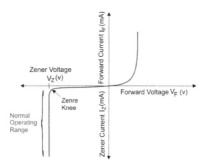

Figure 5.72 V-I characteristics of zener diode.

The above diagram shows the V-I characteristics of a Zener diode. When the diode is connected in forward bias, this diode acts as a normal diode but when the reverse bias voltage is greater than Zener voltage, a sharp breakdown takes place. In the V-I characteristics above V_z is the Zener voltage. It is also the knee voltage because at this point the current increases very rapidly.

- **Ques: 219. Explain the P-N diode characteristics equation?**

Solution:

Let us consider a P-N junction with a donor concentration ND and acceptor concentration NA. Let us also assume that all the donor atoms have donated free electrons and become positive donor ions and all the acceptor atoms have accepted electrons and created corresponding holes and become negative acceptor ions. So, we can say the concentration of free electrons (n) and donor ions ND are the same and similarly, the concentration of holes (p) and acceptor ions (NA) are the same. Here, we have ignored the holes and free electrons created in the semiconductors due to unintentional impurities and defects.

$$n = N_D \quad and \quad p = N_A$$

Across the P-N junction, the free electrons donated by donor atoms in n-type side diffuse to the p-type side and recombine with holes. Similarly, the holes created by acceptor atoms in the p-type side diffuse to the n-type side and recombine with free electrons. After this recombination process, there is a lack of or depletion of charge carriers (free electrons and holes) across the junction. The region across the junction where the free charge carriers get depleted is called the depletion region. Due to the absence of free charge carriers (free electrons and holes), the donor ions of n-type side and acceptor ions of the p-type side across the junction become uncovered. These positive uncovered donor ions towards the n-type side adjacent to the junction and negative uncovered acceptors ions towards the p-type side adjacent to the junction cause a space charge across the P-N junction. The potential developed across the junction due to this space charge is called the diffusion voltage. The diffusion voltage across a P-N junction diode can be expressed as

$$V_D = \frac{kT}{e} \ln \frac{N_A N_D}{n_i^2}$$

The diffusion potential creates a potential barrier for further migration of free electrons from the n-type side to the p-type side and holes from the p-type side to the n-type side. That means diffusion potential prevents charge carriers to cross the junction. This region is highly resistive because of the depletion of free charge carriers in this region. The width of the depletion region depends on the applied bias voltage. The relation between the width of the depletion region and bias voltage can be represented by an equation called the Poisson Equation.

$$W_D = \sqrt{\frac{2\epsilon}{e}(V_D - V)\left(\frac{1}{N_A + \frac{1}{N_D}}\right)}$$

Here, ε is the permittivity of the semiconductor and V is the biasing voltage. So, on an application of a forward bias voltage, the width of the depletion region i.e. P-N junction barrier decreases and ultimately disappears. Hence, in absence of potential barriers across the junction in the forward bias condition free electrons enter into the p-type region and holes enter into the n-type region, where they recombine and release a photon at each recombination. As a result, there will be a forward current flowing through the diode. The current through the PN junction is expressed as:

$$I = I_s \left(e^{\frac{eV}{kT}} - 1 \right)$$

Here, voltage V is applied across the P-N junction and total current I, flows through the pn junction. Is = reverse saturation current, e = charge of the electron, k is Boltzmann constant and T is the temperature in Kelvin scale.

- **Ques: 220. Draw the V-I characteristics of the P-N junction diode?**

Solution:

When V is positive the junction is forward biased, and when V is negative, the junction is reverse biased. When V is negative and less than VTH, the current is minimal. But when V exceeds VTH, the current suddenly becomes very high. The voltage VTH is known as the threshold or cut-in voltage. For Silicon diode VTH = 0.6 V., At a reverse voltage corresponding to point P, there is an abrupt increment in the reverse current. This portion of the characteristics is known as the breakdown region, as shown in Figure 5.73.

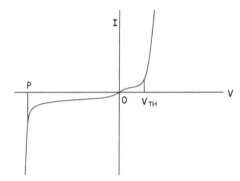

Figure 5.73 V-I characteristics of PN junction diode.

- **Ques: 221. Explain the concept of tunnel diode?**

Solution:

Tunnel diode, as shown in Figure 5.74, is a type of semiconductor diode which is capable of very fast and in microwave frequency range. It was the quantum mechanical effect which is known as tunneling. It is ideal for fast oscillators and receivers for its negative slope characteristics. But it cannot be used in large integrated circuits – that's why it's an application that is limited.

Tunnel diode is one of the most commonly used negative conductance devices. It is also known as Esaki diode after L. Esaki for his work on this effect. This diode is a two-terminal device. The concentration of dopants in both p and n region is very high. It is about 1024 – 1025 m-3 the P-N junction is also abrupt. For such reasons, the depletion layer width is very small. In the current-voltage characteristics of the tunnel diode, we can find a negative slope region when a forward bias is applied.

Quantum mechanical tunneling is responsible for the phenomenon and thus this device is named as tunnel diode. The doping is very high so at absolute zero temperature, the Fermi levels lie within the bias of the semiconductors. When no bias is applied any current flow through the junction.

- **Ques: 222. Explain the characteristics of the tunnel diode?**

Solution:

When a reverse bias is applied the Fermi level of the p-side becomes higher than the Fermi level of the n-side. Hence, the tunneling of electrons from the balance band of the p-side to the conduction band of the n-side takes place. With the interments of the reverse bias, the tunnel current also increases.

When a forward bias is applied to the Fermi level of the n-side becomes higher than the Fermi level of the p-side, thus the tunneling of electrons from the n-side to the p-side takes place. The amount of the tunnel current is very large than the normal junction current. When the forward bias is increased, the tunnel current is increased up to a certain limit.

When the band edge of the n-side is the same as the Fermi level in the p-side, the tunnel current is maximum with the further increment in the forward bias the tunnel current decreases and we get the desired negative conduction region. When the forward bias is raised further, normal PN junction current is obtained which is exponentially proportional to the applied voltage. The V-I characteristics of the tunnel diode are given

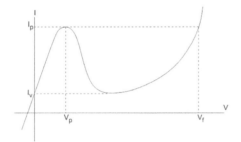

Figure 5.74 V-I characteristics of a tunnel diode.

The negative resistance is used to achieve oscillation and often Ck+ function is of very high-frequency frequencies.

• **Ques: 222. Name any two applications of tunnel diode?**

Solution:

Tunnel diode is a type of semiconductor diode which is capable of very fast and in microwave frequency range. It was the quantum mechanical effect which is known as tunneling. It is ideal for fast oscillators and receivers for its negative slope characteristics. When the voltage is first applied current stars flowing through it. The current increases with the increase in voltage. Once the voltage rises high enough suddenly the current again starts increasing and tunnel diode stars behaving like a normal diode. Because of this unusual behavior, it can be used in several special applications started below:

1. Oscillator circuits

Tunnel diodes can be used as high-frequency oscillators as the transition between the high electrical conductivity is very rapid. They can be used to create oscillation as high as 5Gz. Even they are capable of creativity oscillation up to 100 GHz in an appropriate digital circuit.

2. Microwave circuits

Normal diode transistors do not perform well in microwave operation. So, for microwave generators and amplifiers tunnel diodes are used. In microwave waves and satellite communication equipment, they were used widely, but lately, their usage is decreasing rapidly, as transistors that operate in this frequency range are becoming available.

References

1. http://www.signoffsemi.com/sign-off-checks
2. Millman, J. (1967). *Electronic Devices and Circuits [by] Jacob Millman [and] Christos C. Halkias*. McGraw-Hill.
3. Bogart, T. F., Beasley, J. S., & Rico, G. (2004). *Electronic devices and circuits*. New Jersey: Pearson/Prentice Hall.
4. Gayakwad, R. A., & Gayakwad, R. A. (1988). *Op-amps and linear integrated circuits* (Vol. 25). Englewood Cliffs: Prentice-Hall.
5. https://www.electrical4u.com/diode-working-principle-and-types-of-diode/

Annexure I: Digital Circuit IC Numbers

S. Nos.	Digital Logic	Parameters	IC/Board Nos.
1	Logic gates	Quad 2-input AND logic gate	7408
2		Quad 2-input OR logic gate	7432
3		NOT logic gate/hex inverter	7404
4		Quad 2-input NAND logic gate	7400
5		Quad 2-input NOR logic gate	7402
6		Quad 2-input Exclusive OR logic gate	7486
7		Quad 2-input Exclusive NOR logic gate	74266 (TTL) 4077 (CMOS)
8	Multiplexer	2:1 Multiplexer	74157
9		4:1 Multiplexer	74153
10		8:1 Multiplexer	74151
11		16:1 Multiplexer	74150
12	Demultiplexer	1:2 Demultiplexer	74LVC1G19
13		1:4 Demultiplexer	74139
14		1:8 Demultiplexer	74138
15		1:16 Demultiplexer	74154
16	Decoder	2: 4 Decoder	74155 (TTL)
17		3:8 Decoder	74137/74138
18		4:16 Decoder	74154
19		BCD to decimal decoder	7441
20		BCD to seven segment decoders	7446/7447
21	Encoder	8:3 Priority Encoder	74148
22		10:4 Priority Encoder	74147

S. Nos.	Digital Logic	Parameters	IC/Board Nos.
23	Digital Comparator	4-bit magnitude Comparator	7485
24		8-bit magnitude Comparator	74682
25	Flip-flop	SR Flip-flop	74279
26		JK Flip-flop	7470
27		JK Master Slave Flip-flop	7471
28		D Flip-flop	7474/7479
29		T Flip-flop	7473 short J & K
30	Shift register	8-bit Serial-In-Serial-Out (SISO) register	7491
31		8-bit Serial-In-Parallel-Out (SIPO) register	74164
32		16-bit Parallel-in-Serial-Out (PISO) register	74674
33		4-bit Parallel-in-Parallel-Out (PIPO) register	7495
34	ADC and DAC	16-bit A/D converter	ADS5482 (TI)
35		16- bit D/A converter	DAC8728 (TI)
36	Adder and Subtractor	2-bit Full Adder	7482
37		4-bit Full Adder	7483
38		4-bit Full Subtractor	74385
39	Counter	Up-down binary counter	74191
40		Up-down decade counter	74190
41		Modulo 10 counters	74416
42	Programmable Logic Devices (PLD)	Field Programmable Gate Array (FPGA)	SPARTAN 6 family, ARTIX 7 family
43		Complex Programmable Logic Device (CPLD)	ALTERA MAX 7000 series
44	Memories	16-bit RAM	7481/7484
45		64-bit RAM	7489
46		256-bit ROM	7488
47		512-bit ROM	74186 (open collector)

S. Nos.	Digital Logic	Parameters	IC/Board Nos.
48		256-bit PROM with open collector output	74188
49		2048-bit PROM with open collector output	74470
50		2048-bit PROM with three state output	74471
51		1024-bit PROM with three state output	74287

Annexure II: List of Keywords, System Tasks, and Compiler Directives Used in Verilog HDL

This list consists of keywords, system task, and compiler directives. All the keywords are defined in lowercase. The system tasks are *tasks* and functions that are used to generate input and output during simulation. The compiler directives are used to control the compilation of a *Verilog* description. The reference is IEEE std. 1364-2001, Verilog HDL.

Keywords		System Tasks	Compiler Directives
always	module	$bitstoreal	'accelerate
assign	task	$countdrivers	'autoexpand_vectornets
begin	library	$display	'celldefine
fork	time	$fclose	'default_nettype
case	table	$fdisplay	'define
casex	specify	$fmonitor	'define
casez	join	$fopen	'else
buf	end	$fstrobe	'elsif
bufif0	endcase	$fwrite	'endcelldefine
bufif1	endtable	$finish	'endif
rtran	endprimitive	$getpattern	'endprotect
rtranif0	endmodule	$history	'endprotected
rtranif1	endspecify	$incsave	'expand_vectornets
defparam	endtask	$input	'ifdef
deassign	pull0	$itor	'ifndef
include	pull1	$key	'include
integer	pullup	$list	'noaccelerate

Keywords		System Tasks	Compiler Directives
instance	pulldown	$log	'noexpand_vectornets
automatic	tri	$monitor	'noremove_gatenames
cell	tri0	$monitoroff	'nounconnected_drive
cmos	tri1	$monitoron	'protect
pmos	force	$nokey	'protected
nmos	forever	$stop	'remove_gatenames
and	real	$finish	'remove_netnames
or	reg	$write	'resetall
not	repeat	$rtoi	'timescale
nand	if	$readmemh	'unconnected_drive
nor	else	$readmemb	'undef
strong0	parameter	$hold	
strong1	primitive	$period	
supply0	wait	$skew	
supply1	wire	$timeformat	

Index

About the Authors

Dr. Cherry Bhargava is working as an associate professor and head, VLSI domain, School of Electrical and Electronics Engineering at Lovely Professional University, Punjab, India. She has more than 15 years of teaching and research experience. She is PhD (ECE), IKGPTU, M.Tech (VLSI Design & CAD) Thapar University and B.Tech (Electronics and Instrumentation) from Kurukshetra University. She is GATE qualified with All India Rank 428.

She has authored about 50 technical research papers in the SCI, Scopus indexed quality journals and national/international conferences. She has 12 books related to reliability, artificial intelligence and digital electronics to her credit. She has registered three copyrights and filed 20 patents. She is the recipient of various national and international awards for being an outstanding faculty in engineering and an excellent researcher. She is an active reviewer and editorial member of various prominent SCI and Scopus indexed journals. She is a lifetime member of IET, IAENG, NSPE, IAOP, WASET and reliability research group. Her area of expertise includes the reliability of electronic systems, digital electronics, VLSI design, artificial intelligence, and related technologies.

Dr. Gaurav Mani Khanal is working as a Post-doctoral researcher in DSP-VLSI lab, Department of Electronics Engineering, University of Rome Tor Vergata. Rome (Italy). He earned his PhD in Electronics Engineering (Memristor and Memristive System design and Fabrication) from the University of Rome Tor Vergata, Rome (Italy), Rome (Italy). He has obtained Advanced Master's Degree in Wireless Systems and Related Technology from Politechnico Di Torino (Polytechnic University of Turin), Turin (Italy). Dr. Gaurav has completed a Master of Science (M.S) in Microelectronics and System Design from Liverpool John Moores University, Liverpool (United Kingdom).

He possesses excellent knowledge of 2D material especially ZnO-Graphene and their composite based low power semiconductor electronics devices (theory and fabrication), memristor/resistive switching device modelling, fabrication (spin and dip coating) and characterization (XRD, SEM, EIS). He has working experience with tools like Pspice, MATLAB, ModelSim, Tanner and Knowledge Verilog HDL for digital system design. He had hands-on experience of application development with Arduino, Xilinx Spartan FPGA and R-pi boards.